Through the Looking Glass...

A Field Guide to Aquatic Plants

"*Oh, how nice it would be if we could only get through into the Looking Glass House! I'm sure it's got, oh such beautiful things in it! Let's pretend there's a way of getting through into it somehow. Let's pretend the glass has got all soft like gauze, so that we can get through. Why, it's turning into a sort of mist now, I declare! It'll be easy enough to get through...*"

From **Through the Looking Glass**, Lewis Caroll, 1862

Text and Layout, ©1997 Wisconsin Lakes Partnership

Illustrations 1997 Carol Watkins, Horseshoe Hollow Graphics

ISBN 0-932310-32-X

DNR Publication # FH-207-97

Please direct comments or inquiries to:

Wisconsin Lakes Partnership
University of Wisconsin–Extension
University of Wisconsin–Stevens Point
College of Natural Resources
1900 Franklin St.
Stevens Point, WI 54481
715/346-2116
Fax: 715/346-4038

Printed on
Recycled Paper

Printing by Reindl Printing, Inc. • Merrill, WI

Authors
Susan Borman, Robert Korth, Jo Temte

Illustrations
Carol Watkins

Production Editor
Dorothy Snyder

Graphics Design
Jeffrey Strobel

This guidebook has been made possible because of the willingness of many people to share their expertise and time. We hear much about the philosophy of collaboration and partnerships. *Through the Looking Glass* is an example of that philosophy at work.

Contributors
Dan Helsel
Chris Hinz
Brad Johnson
Joanne Kline
Deb Konkel
Celeste Moen
Charmaine Robaidek
Barb Timmel

Technical Assistance and Review
Stan Nichols
Sandy Engel
Robert Freckmann
Don Reed
Tamara Dudiak
Sara Rogers

Special thanks to:

Carol Watkins... for the hundreds of painstaking hours spent fashioning her love for nature into the beautiful art that graces the pages of this guide. Working from real specimens, Carol meticulously created these incredibly detailed and accurate illustrations that are the essence of this text.

Environmental Resources Center, University of Wisconsin–Extension, Madison... for their assistance with this manuscript.

We also thank the following individuals for their advice and assistance: Art Bernhardt, Jeff Bode, Laura Herman, Frank Koshere, Jim Leverance, Dave Marshall, Tim Rasman, Dan Ryan, Mark Sesing, Buzz Sorge, Scott Szymanski, Bob Wakeman, Bob Young.

Contents

Emergent Plants (Plants with leaves that extend above the water surface.)

Plants with Broad Leaves

Free-floating Plants (Plants that float freely on the water surface.)

Floating-leaf Plants (Plants with leaves that float on the water surface.)

Submersed Plants (Plants with most of their leaves growing below the water surface.)

Plants with Entire Leaves – Opposite or Whorled

Plants with Entire Leaves – Alternate or Basal

Plants with Finely-Divided Leaves

Foreword

The real magic of aquatic plants can be observed by slipping *through the looking glass.* In this world beneath the waves, plants unequipped to withstand the force of gravity are buoyed up in a delicate splendor. This underwater forest is a place where dream and reality move side by side.

As on land, creatures exist at all levels in these aquatic groves. Small fishes soar like birds around the plants. Larger predators lurk in their shadows. Insects, snails, bryozoans, sponges and other curious creatures live out their secret lives in the nooks and crannies of this muted forest.

This is a special place moving to an ancient rhythm. All manner of creatures come together from the land, air and water. Through this text we hope to acquaint you with the world of aquatic plants. If you decide to venture *through the looking glass,* perhaps this guide will add to your understanding and increase the richness of your experience.

R. Korth

A boat, beneath a sunny sky
   Lingering onward dreamily
   In an evening of July —

Children three that nestle near,
   Eager eye and willing ear,
   Pleased a simple tale to hear —

Long has paled that sunny sky:
   Echoes fade and memories die:
   Autumn frosts have slain July.

Still she haunts me, phantomwise.
   Alice moving under skies
   Never seen by waking eyes.

Children yet, the tale to hear,
   Eager eye and willing ear,
   Lovingly shall nestle near.

In a Wonderland they lie,
   Dreaming as the days go by,
   Dreaming as the summers die:

Ever drifting down the stream —
   Lingering in the golden gleam —
   Life, what is it but a dream?

Note: The first letter of each line when read downward spell the name of Alice Pleasance Liddell, the little girl for whom the story, *Through the Looking Glass,* was originally told.

Introduction

The Guide

We hope this text will increase your interest and awareness of aquatic plants and their relationship to a sound lake environment. The narrative concentrates on natural history and provides precise information and details to help you identify aquatic plants. The idea is to make learning about these plants fun and exciting by providing an easy-to-use field guide.

The selection of plants included in the guide was based partly on the Wisconsin Lake Plant Atlas, written by Stan Nichols. While the guide focuses on plants found in Wisconsin waters, they are also found in waters across the United States and Canada, and a number of them have global distribution.

Studying Aquatic Plants

Imagine a plant that can grow entirely underwater. Many aquatic plants must be able to grow with up to 95% less light than their land-living counterparts. Long periods with waterlogged roots kill most land plants, but submersed aquatic plants can exchange gases and nutrients in a saturated world.

Aquatic plants can be harder to identify than land plants. They are basically green and few have conspicuous flowers. Their shapes and size can vary with growing conditions. Add to that the complication of getting close to a plant that is in, or under, water, and . . . well, you get the picture.

To get a good look at aquatic plants you must explore wet places. If a canoe or boat won't work, you will need to proceed on foot. These places are truly crawling with life. Come prepared for soft bottoms and bugs. You will uncover the home of birds, butterflies, frogs and fish, to mention only a few.

We believe that the extra effort to study aquatic plants is worth the reward. If you slow down and look closely, you will find a place full of surprise and delight!

SETTING THE SCENE

A *close look at a lake reveals a dynamic community. A host of characters such as fish, birds, mammals, snails, crayfish, insect larvae and tiny hydras all interact with plants on a submersed stage. In many ways these underwater meadows set the scene and determine the script for the timeless drama that unfolds in the shallow-water environment. As you investigate your lake's nearshore area, you will discover a variety of plant communities that gradually change as the water gets deeper. This section examines these linkages.*

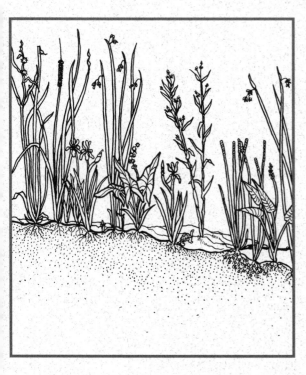

The Emergent Plant Community

Let's wade right in. In the shallowest zone, from moist shoreline soils to knee-deep water, grow the emergent plants of the shallow-water community. There are sword-like blades of cattail, bur-reed and blue flag iris, clumps of bottle brush sedge and cylindrical stalks of bulrush and spikerush. Look also for the arrow-shaped leaves of duck potato and heart-shaped foliage of pickerelweed.

Emergent plants can tolerate fluctuating water levels and their dense stands can dampen shoreline waves. Like terrestrial plants, the leaves of the emergents have a protective waxy coating, called a cuticle, that helps retain moisture. However, emergents differ from upland plants in a number of important ways.

- The leaves have extensive spongy tissue, called aerenchyma, and air spaces. This makes them great nesting material for ducks, shorebirds and muskrats. Nests made of these buoyant leaves float up and down with changing water levels.

- The roots spread horizontally creating an interlocking network like a jute-backed carpet. This growth pattern is very important for stabilizing sediment. It also helps these plants withstand wave action and dissipate the force of upland runoff.

- Flexible reproductive strategies allow emergents to take advantage of variable conditions. When water levels are low, they reproduce from seeds that germinate on exposed mud flats. When water levels are high, they are equally successful at staking out territory with spreading roots and horizontal buried stems, called rhizomes, that send up new shoots.

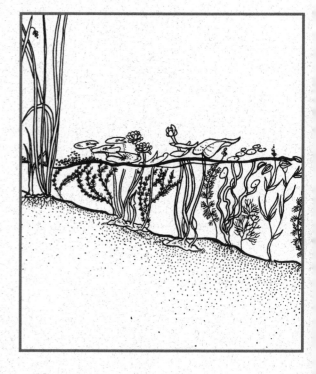

The Floating-Leaf Plant Community

As you slosh from water knee-deep to water up to your waist and deeper, you'll notice that the plant community changes. Cattails and bulrushes gradually give way to floating-leaf plants and a variety of submersed species. White and yellow water lilies, watershield, floating-leaf pondweeds and duckweed crown the surface. Strands of submersed plants such as coontail, waterweed, bladderwort and water milfoil spread out beneath this canopy.

To adapt to life in this zone, floating-leaf plants have developed special features.

- Did you know that some water lily beds are hundreds of years old? Extensive rhizomes that expand every growing season are very long-lived. These rhizome networks are important sediment stabilizers.

- Circular or elliptical leaves with smooth margins help resist tearing by wind or waves. The leathery texture of the leaves also makes them durable. A thick cuticle protects the upper leaf surface, reduces moisture loss, and sheds water to help prevent immersion. With a magnifying lens, you can see small holes for air exchange, called stomata, spread over the upper leaf surface. Cut a leaf open to discover a network of air spaces that stores gases and helps leaves float.

- Slack and elastic leaf stalks, called petioles, help floating leaves ride the waves and allow flexibility of leaf arrangement on the surface. Sometimes the petiole is bent just below the point of attachment to the leaf; this helps the leaf blade lie horizontally on the water surface. Petioles of floating leaves can grow rapidly. In the spring, rapid growth allows the leaf to reach light for photosynthesis before energy stored in the rhizomes is depleted. This capacity for rapid growth is also important when flooding immerses the plant.

- These plants have developed specialized reproductive and overwintering strategies. Rooted floating-leaf plants die back to the rhizomes in winter and send up new leaves in the spring. They can also reproduce from seeds.

- Some plants in this community are free-floating. The duckweeds, watermeals, and riccias are completely unattached and drift with the wind and current. These plants look like small flattened leaves. Roots are either absent or simple hair-like projections dangling from the underside of the leaf. Free-floating plants form modified leaf structures called turions that sink in the fall and overwinter on the bottom of the lake. In the spring they become buoyant and float to the surface where they produce new leaf growth.

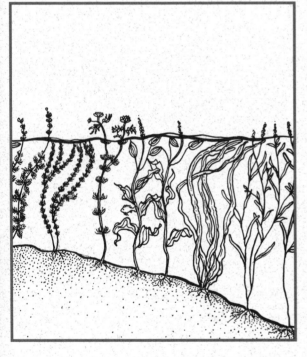

The Open Water Submersed Plant Community

Now we'll slip deeper through the looking glass. Open water plants are most often seen wrapped on the end of an anchor or fishing line. You'll need to start snorkeling or rig up a glass-bottom view scope to appreciate the diversity of these plants. Beyond the zones of shoreline emergents and lily beds lies the primary domain of submersed plants. These underwater plants grow in water ranging from the shallowest zone to the cool depths of several meters. Their maximum depth is limited only by available light.

The diversity of form and structure that can be found in this submersed community is impressive. Ribbon-like strands of wild celery, feathery leaves of water milfoil, coontail, bladderwort and water marigold, broad arching leaves of large-leaf pondweed and stiff linear leaves of flat-stem pondweed intertwine in the lake's depths. A variety of growth forms and life cycles make this a constantly changing aquascape.

Life underwater demands special adaptations. These submersed plants have evolved furthest from their terrestrial ancestors.

- Submersed leaves are flexible and often finely divided, decreasing resistance to water movements. This intricate shape provides excellent habitat for invertebrates with ample surface area and plenty of places to trap bits of food.

- Since water conservation is no problem, the leaves lack cuticle. The absence of cuticle facilitates the exchange of gases between the plant and water. The spongy interior of the leaf helps with gas storage and flotation. Leaves must be buoyant rather than limp to capture light filtering down from the lake surface.

- Many submersed plants have more than one type of leaf on the same plant – a condition called heterophylly. Most often these plants have a combination of submersed and floating leaves or submersed and aerial leaves. Floating or aerial leaves are usually less divided, thicker and protected by cuticle.

- Stems containing chlorophyll act like leaves to help submersed plants capture light for photosynthesis. Stems and leaves share other characteristic as well, including flexibility and interior spongy tissue for buoyancy.

- The height of submersed plants can vary dramatically depending on available light, nutrients and where they take root. The same species may be ankle-high in shallow water and taller than a person in deep water.

- A diversity of root systems allows a variety of submersed plants to grow in close proximity. Some roots are shallow and spreading; others are deeper and tuberous.

- These plants have a variety of reproductive and overwintering strategies. Like emergent and floating-leaf plants, submersed plants rely heavily on vegetative means for expansion and overwintering. Submersed plants may produce rhizomes or tubers (wild celery, pondweeds), winter buds or turions (pondweeds, bladderworts, duckweeds), or overwinter as a whole plant (waterweed, coontail, curly-leaf pondweed).

- Some submersed plants, including naiads and horned pondweed, do rely on sexual reproduction for survival. These are the true annuals. They produce abundant seeds in fall and seedlings in spring. Most other submersed plants rely mainly on clonal expansion.

Scripting and Casting

The character of the shallow water community is determined by the diversity, distribution and density of aquatic plants. Let's take a look at your lake's potential cast of critters and the role plants play in their lives.

Invertebrates

One of the important links in any aquatic ecosystem is the connection between plants and invertebrates. Filter-feeders attach to plants and take their food from the surrounding water. Insect larvae and nymphs cling to stems and hide among leaves as they search for prey. Algae and diatoms on plant surfaces are grazed by snails and midges. Caddisfly and moth larvae feed directly on plant tissue. Many invertebrates rely on plants during specific life stages. Eggs are deposited on leaves and stems and larvae burrow into stems or wrap themselves in leaves for the transformation to adult forms.

A diverse community of plants can support a wide range of invertebrates. Feeding needs, life cycle requirements and predator-prey relationships are built around the presence of specific plants. Sediment dwelling invertebrates can also benefit from the presence of plants. Plants stabilize the sediment, reduce erosion, buffer currents and enrich the sediment with detritus.

Fish

Any angler knows how important plants are to fish. Habitat created by aquatic plants provides food and shelter for both young and adult fish. Invertebrates living on or beneath plants are a primary food source. Some fish, particularly bluegills, also graze directly on leaves and stems.

Predatory fish cruise the shadowy plant beds in search of prey. The structure and density of the plant beds can determine whether predators will be well-fed or go hungry. Too few plants can limit the number of prey fish; plant growth that's too dense can fence predators out. Scientists estimate the optimal plant cover for northern pike to be greater than 80%. Yellow perch are most successful at 25-50% plant cover, for bluegills 15-30% and for largemouth bass 40-60%.

Bass and bluegills use shallow plant beds for spawning. They clear the immediate nest site, but the surrounding vegetation helps buffer wave action and is important cover for larval and juvenile fish. Northern pike also seek shoreline vegetation for spawning. Young-of-year pike success is much higher in vegetated areas.

Waterfowl and Shorebirds

The significance of aquatic plants for waterbirds is often underestimated. Plants offer food, shelter and nesting material. A diversity of plants provides food throughout the seasons. Waterfowl and shorebirds also eat the invertebrates that live on these plants.

Many ducks make seasonal diet changes. Breeding hens switch from foliage to invertebrates before laying their eggs. Ducklings move from a diet rich in invertebrates when they are young to seeds, tubers and shoots as they get older. Migratory ducks, such as canvasbacks, rely on a high-energy carbohydrate diet of wild celery and sago pondweed tubers during fall migration.

Shoreline emergents provide camouflage and protection from wave action for young waterfowl. Their buoyant leaves also make ideal nesting material. Common loons use available plant matter to build their mounded nests on the shoreline.

Mammals

It's not unusual to see a deer or moose standing in shallow water with a big mouthful of pondweed. Otter families patrol the plant beds hunting their dinner. Cruising nearby may be a muskrat with a cattail in its mouth or a beaver diving down to munch on a water lily tuber. The plants of the nearshore area are especially important for these shoreland mammals.

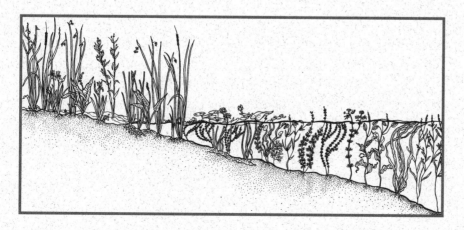

A Plant Worth Having

Aquatic plants can influence lake characteristics that increase the value of aquatic habitats for both human and wildlife uses.

A Nuisance?

Aquatic plants are an indispensable part of a lake ecosystem. Nevertheless, some lakefront property owners are frustrated by dense beds of aquatic plants and consider them a nuisance.

Defining a nuisance based upon its impacts to human activities is difficult. Each lake user has a different tolerance of aquatic plant densities. Anglers correlate lush aquatic plants with good fishing. Duck hunters and wildlife watchers also consider aquatic plants beneficial. People who grew up swimming in lakes with moderate densities of aquatic plants don't mind swimming through fine stems of pondweed or wading on top of a mat of chara. However, swimming through a dense stand of Eurasian water milfoil can create an unpleasant situation for even the most tolerant swimmers.

Boaters consider filamentous algae or floating vegetation a nuisance when it's thick enough to clog propellers. Drifting aquatic plants can cover shore frontages and create unpleasant odors as they decay. This can bring beach use to a halt.

Nuisance aquatic plant conditions are typically an indication of bigger problems. Shallow, nutrient-rich lakes with plenty of cultural disturbance in their watershed are great candidates for heavy plant growth. Activities that remove shoreline vegetation and expose soil – such as construction, logging and agriculture – allow sediments to move into the water, creating a source of nutrients for plant growth. Leaking septic systems and lawn fertilizers can add even more nutrients.

It's important to consider the benefits of aquatic plants as well as the potential cost to recreational use. Protecting and enhancing native plant populations makes good sense, both for the lake ecosystem and the value of the surrounding property. Directing well-planned management toward nuisance aquatic plant conditions also makes good environmental and financial sense.

Water Quality

Aquatic plants can improve water quality. They absorb phosphorus, nitrogen and other nutrients from the water that could otherwise fuel nuisance algal growth. Some plants can even filter and break down pollutants. Finely divided foliage acts as a filtering system that traps and settles particles from upland runoff. Resuspension of sediment is also lessened by the interlocking network of plant roots and rhizomes.

When plant growth becomes too dense, dramatic daily shifts in dissolved oxygen and pH may occur. During the winter, dense stands of decaying plants can lead to low dissolved oxygen conditions under the ice. In general, these nuisance conditions are not observed in healthy, diverse stands of native aquatic plants. More often than not, plants associated with these impacts are exotic species like Eurasian water milfoil or curly-leaf pondweed. A well-designed aquatic plant management plan addresses these problems and helps restore water quality and a healthy balance in the aquatic community.

Shoreline Protection

Stands of emergent plants and flotillas of water lilies blunt wave action and buffer the shoreline. By stabilizing the land and water interface, sediment and debris are captured and erosion is reduced.

Beauty and Value

One of the immeasurable benefits of aquatic plants is the beauty they contribute. Who can put a value on the purple spires of pickerelweed, delicate white blooms of duck potato or the snapdragon-like flowers of bladderwort floating just above the surface? We are gradually realizing that one of the greatest resources of our region is its natural beauty – and the native aquatic plants of our lakes, streams and rivers are an important part of this resource.

So there you have it! The scene is set…a cast of marvelous characters, a spectacular stage and a wondrous plot. Add a pinch of your curiosity and you have a recipe for a journey of satisfying discovery!

USING THE GUIDE

Through the Looking Glass is designed as an easy-to-use guide, not a taxonomic key. This segment describes the purpose of the various icons, categories and terms used throughout the guide.

The Major Plant Sections

To make it easier to find the plant you are looking for, the plant descriptions are divided into four major sections. Each section corresponds to where the plants normally grow and the plants' shape.

Emergent Plants have leaves that extend above the water's surface and are usually found in shallow water. Some of the smaller, narrow-leaved emergents also grow submersed in shallow water.

They are subdivided into:

>   **Narrow-leaved** (cattail, iris)

>   **Broad-leaved** (arrowhead, pickerelweed)

Free-floating Plants are not attached to the bottom by roots or rhizomes (duckweed, watermeal). These plants are small and can be found floating anywhere on the water's surface, depending on wind and currents.

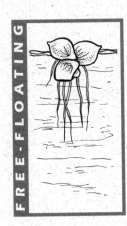

Floating-leaf Plants are rooted plants with leaves that float on the surface (water lily, lotus). They are usually found at intermediate depths between the shallow emergent zone and deeper submersed plant beds.

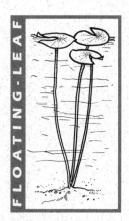

Submersed Plants have most of their leaves growing underwater; some floating leaves may also be present. They are found from shallow to deep zones. Some of them have flower stalks that stick up out of the water.

They are subdivided into:

Entire Leaves – Opposite or Whorled
(naiad, waterweed)

Entire Leaves – Alternate or Basal
(pondweed, wild celery)

Finely-divided Leaves (coontail, milfoil)

Plant Names

Within the sections, plant descriptions are arranged alphabetically by scientific name. The most frequently used common name for each plant is listed first at the top of the page and in the table of contents. If plants have more than one common name, they are also listed. All plants are listed in the index under both common and scientific names.

Common names are drawn from a number of sources. Scientific names come from a commonly accepted authority, *A Manual of Vascular Plants of Northeastern United States and Adjacent Canada,* 2nd ed. by Henry Gleason and Arthur Cronquist. Occasionally a scientific name is changed based on new information about the plant. For plants that have recently undergone a name change, we reference the old name (formerly known as...). For plants that are currently undergoing a name change, we provide other current names (also known as...). A phonetic pronunciation is given for each scientific name. We have also included a definition of the scientific names.

In the index we have listed the scientific name (such as *Myriophyllum farwellii* Morong) with what is known as a naming authority. The authority, like Morong, is the person or persons who named the species. The names of the authorities are often abbreviated in a standardized way. The most common example is "L." for Linnaeus. The complete scientific name (including authority) is used in scientific reports and on plant specimens collected for herbariums.

In a few cases, we have described a genus as a whole. In these instances, the genus name is given followed by "spp." – which means more than one species. This was done where the genus is well-known, but several similar species are more easily discussed as a group.

Icons

Visual images are easy to locate and provide a fast way to recognize various plant features. Wherever possible, icons were used to denote special features, concerns or categories.

The boxed icons provide a quick reference about plant features. The upper left box shows the **plant category** (Emergent, Free-floating, Floating-leaf, Submersed).

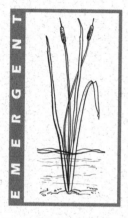

Further divisions of plant sections, for example **Emergent Plants with Broad Leaves** and **Emergent Plants with Narrow Leaves**, are noted above the symbol for the plant section where applicable.

The lower right box shows the **plant's place of origin –** whether the plant is **N**ative or **E**xotic and if it is listed as **R**are in Wisconsin.

The Wisconsin Department of Natural Resources (DNR) Bureau of Endangered Resources lists rare plants in the following categories:

Endangered – Those species whose continued existence as a viable component of the state's wild plants is in jeopardy.

Threatened – Those species which appear likely to become endangered.

Special Concern – Those species about which some problem of abundance or distribution is suspected but not yet proved. The main purpose of this category is to focus attention on certain species before they become threatened or endangered.

Plant Facts

Each aquatic plant described in this guide is accompanied by information specific to the species. These details are broken into seven areas:

Description lists distinguishable parts and overall appearance which will help you recognize the plants. Measurements are indicated in metric units to be compatible with other plant guides. You will find a metric ruler and conversion chart on page 248 for easy reference.

Some of the key features have been enlarged and highlighted in the illustrations. These original drawings were based on actual plants, herbarium specimens and technical guides.

An effort has been made to limit scientific jargon. If a term is unfamiliar, check the illustrated glossary at the back of the book.

Similar Species describes unique features to help you distinguish this plant from others. Additional information about some of the related plants is also provided. Plants that are only described as similar species, and don't have their own description page, are highlighted in bold type.

Origin and Range tells if a plant is native or exotic, gives a general description of its distribution in Wisconsin, and includes its range within the United States. Many of these plants are also found in Canada, Central America or even globally.

Habitat describes the best conditions for growth of this plant and tells where to find it.

Through the Year shows the yearly growth cycle of the plant. Understanding growth cycles can be a powerful tool in determining why a plant expands or declines and how it can best be protected or controlled.

Value in the Aquatic Community explains the worth of a plant species to the other members of its ecosystem. These details can be used for improving and restoring habitat, limiting erosion, and providing food and shelter for a variety of critters.

A Closer Look investigates the world of aquatic plants that is often hidden from view. Learning about exceptional or unusual features can pique your curiosity. Uncovering a plant's remarkable background can raise your appreciation.

Sources

The information in this guide is drawn from many sources. The descriptions and technical details are based on a blend of the best identification tips given in regional technical guides and from actual observations made in the field. If you want to find out more about topics that interest you, references and appropriate credits are given in the back of the book.

The size of various plant parts is based primarily on measurements given by Henry Gleason and Art Cronquist in *A Manual of Vascular Plants of Northeastern United States and Adjacent Canada,* 2nd ed. This information was verified with other guides and herbarium specimens.

Distribution and range is supported by data from the *Wisconsin Lake Plant Atlas* by Stan Nichols, the *Guide to Wisconsin's Endangered and Threatened Plants* by Wisconsin DNR, and the Henry Gleason and Art Cronquist manual.

The particulars for habitat requirements and seasonal cycles are based on information in guides written by Steve Eggers, Donald Reed, Sandy Engel, Stan Nichols, Edward Voss and Norman Fassett (see bibliography). Other details come from journal articles and actual field observations.

Emergent Plants

> "Hope and the future for me are not in lawns and cultivated fields, not in towns and cities, but in the impervious and quaking swamps."
>
> Henry David Thoreau, 1851

Plants with Narrow Leaves

Sweetflag
Flowering rush
Sedges
Bristly sedge
Three-way sedge
Needle spikerush
Creeping spikerush
Robbins spikerush
Water horsetail
Northern manna grass
Northern blue flag
Soft rush

Brown-fruited rush
Rice cut-grass
Reed canary grass
Giant reed
Hardstem bulrush
Three-square
River bulrush
Softstem bulrush
Common bur-reed
Narrow-leaved cattail
Broad-leaved cattail
Wild rice

Plants with Broad Leaves

Water plantains
Wild calla
Water hemlock
Swamp loosestrife
Purple loosestrife

Pickerelweed
Marsh cinquefoil
Water cress
Grass-leaved arrowhead
Common arrowhead

Acorus calamus (AK-or-us CAL-a-mus)

Sweetflag, calamus

Acorus – (L.) aromatic plant; *calamus* – (Gk.) *kalamos:* reed

At first glance it looked like cattail. When the woman brushed against this relative of jack-in-the-pulpit, the leaves released a fresh, pleasingly spicy fragrance that reminded her of vanilla.

Description: Sweetflag has tall, sword-like leaves (1.2-2 m high, 8-25 mm wide) that are crowded at the base where they emerge from the rhizome. The flower stalk resembles the leaves but has a finger-like flower spike called a spadix that juts out at an angle. Tiny yellowish-brown flowers are crowded on the spadix (5-10 cm long, 1-2 cm thick).

Similar species: The leaves of sweetflag are similar in size and shape to cattail (*Typha* sp.) or bur-reed (*Sparganium* sp.). However, the leaves of sweetflag have one wavy margin and an off-center midrib. Sweetflag can also be recognized by the "sniff test." All portions of the plant release a sweet, spicy smell when crushed.

Origin & Range: Native; scattered locations throughout Wisconsin; range includes most of U.S.

Habitat: Sweetflag is found on the margins of lakes and ponds and in marshes and wet meadows. It is usually limited to water less than 0.5 meter deep.

Through the Year: Sweetflag is perennial, resprouting each spring from the rhizome buried in the sediment. There is also limited reproduction from seed. The tiny flowers bloom from early to midsummer. By late summer the spadix is covered with small fruit that gradually ripen.

Value in the Aquatic Community: The leaves of sweetflag provide cover and nesting material for waterfowl and shorebirds. It is also eaten by muskrats. Shoreline stands of sweetflag can be important for stabilizing sediment, buffering wave action and preventing erosion.

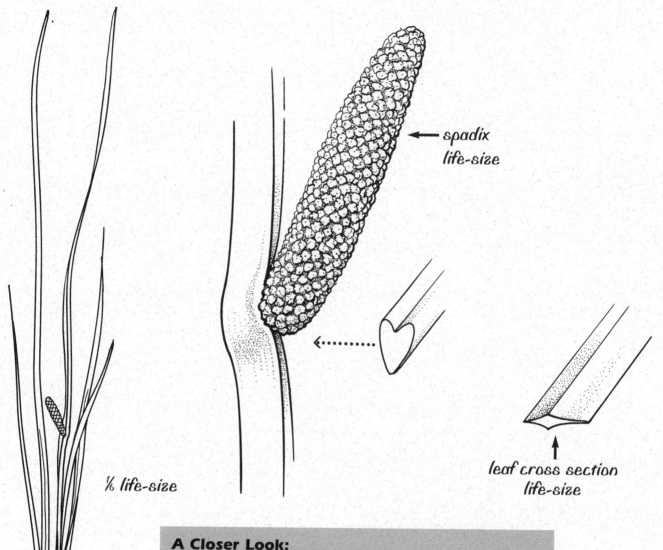

spadix
life-size

⅙ life-size

leaf cross section
life-size

A Closer Look:

Humans have a rich and ancient relationship with sweetflag. Genetic studies have shown that it may be the oldest surviving descendent of ancestral monocots (Duvall 1993). People have been using sweetflag for over 4,000 years for culinary and medicinal uses. The roots have been candied and sold as a confection. Oil of calamus, derived from the leaves and roots, is very aromatic. It has been used in perfumes, snuff and flavored cordials including Benedictine and Chartreuse.

There is an active ingredient called Beta-asarone in the root extract that acts as an insecticide, particularly against biting and sucking insects. Calamus has also been used in herbal remedies ranging from painkillers to cough medicine. Currently calamus is classified as an unsafe herb by the Food and Drug Administration (FDA). Long-term feeding studies with rats have shown oil of calamus to have a variety of toxic effects. It is suspected that asarone is one of the components that creates this toxicity (Duke 1985).

Native

EMERGENT

Butomus umbellatus (BEW-toe-mus um-bell-LAY-tus)

Flowering rush

Butomus – (Gk.) *boutomos* – sedge; *umbellatus* – (L.) *umbella:* parasol

*A flash of pink along the shoreline reveals the invader.
A cluster of pink blossoms sways in the breeze atop a baton-
like stalk. Beneath the surface the exotic interloper floats as
long, trailing leaves without the colorful warning flag.*

Description: Flowering rush has linear leaves (1 m tall, 5-10 mm wide) that emerge from a stout rhizome. These leaves may be emergent, floating or even submersed in water up to several meters deep. Submersed leaves are long and tape-like, resembling wild celery but without the central stripe. The flower stalk rises above the leaves (1-1.5 m) with a terminal cluster of many pink flowers. Flowers (2-2.5 cm wide) are on slender stalks (5-10 cm long) and have three sepals and three petals. Fruit develops as a cluster of follicles with long beaks.

Similar species: Flowering rush could be confused with some of the narrow-leaved water plantains (*Alisma* spp.) or arrowheads (*Sagittaria* spp.). However, these species don't have pink flowers or the narrow, grass-like leaves of flowering rush. Leaves of sterile plants might be confused with wild celery (*Vallisneria americana*), bur-reeds

(*Sparganium* spp.) or arrowheads. The sterile leaves of flowering rush lack the prominent median stripe of wild celery and the cross veins evident in bur-reed and arrowhead.

Origin & Range: Exotic; originated in Europe and temperate areas of Asia. Distribution in Wisconsin is not well-documented. Flowering rush was introduced in the St. Lawrence River area and has spread from there. New sites have developed in a number of states including Illinois, Minnesota, Wisconsin and Idaho.

Habitat: Flowering rush grows on shorelines and marshes as well as submersed in lakes and streams. It will grow in a variety of sediments at depths ranging from very shallow to several meters deep.

Through the Year: Flowering rush is a perennial, resprouting each spring from the winter-hardy rhizome. Reproduction from seed is considered rare. It stakes out new territory through growth of the rhizome and dense stands can develop. Flowering occurs midsummer and fruit is set by late summer.

Value in the Aquatic Community: Flowering rush may provide cover for waterfowl. However, there is concern that it can crowd out native species with higher wildlife value.

A Closer Look:

Care should be taken to avoid spreading flowering rush. Likely modes of introduction include transport of rhizome fragments and cultivation in water gardens.

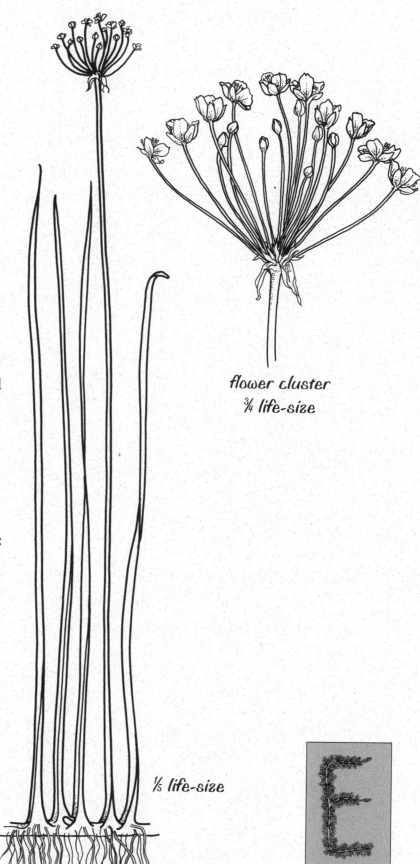

flower cluster
¾ life-size

⅛ life-size

E

Exotic

EMERGENT

Carex spp. (CARE-ex)

Sedges

Carex – (L.) sedge

There is a saying that sedges have edges, referring to their 3-angled stems. It is also true that edges have sedges – the margins of lakes, ponds and streams often have an assortment of Carex species anchoring the shoreline.

Description: The sedges appear grass-like at first glance. However, the leaves come off the stems at three angles as opposed to two angles in grasses. The most definitive characteristic of *Carex* is the sac-like structure, called a perigynium, that surrounds the ovary and nutlet. This is unique to *Carex* and helps separate it not only from grasses but also from other genera in the sedge family. The surface texture, shape and beak of the perigynium are all used in identification.

The tiny flowers of *Carex* are unisexual and each one occurs in the axil of a single scale. Some species have the male and female flowers on separate spikes, while others have them both on the same spike.

Similar species: The genus *Carex* has more species than any other plant genus in Wisconsin. This is true for most temperate climates and makes species identification challenging. *Carex* species can be confused with grasses but the three-ranked leaves and presence of

perigynia help distinguish them.

Origin & Range: Native. There are more than 500 North American species of *Carex*.

Habitat: *Carex* species are found in forests, wet meadows, fens, prairies, bogs and shorelines of lakes, ponds and streams.

Through the Year: Sedge shoots emerge from rhizomes in the spring. Some of these shoots develop in the fall and are dormant over winter. Flower spikes develop by midsummer and the nutlets are mature by late summer.

Value in the Aquatic Community: *Carex* species are an essential food source for a wide variety of wildlife including marsh birds, shorebirds, upland game birds (bobwhite, ruffed grouse, sharp-tailed grouse, pheasant, turkey) and most waterfowl. Sedges also provide food for moose, beaver, deer and muskrat. Stands of *Carex* in shallow water can also provide valuable spawning habitat.

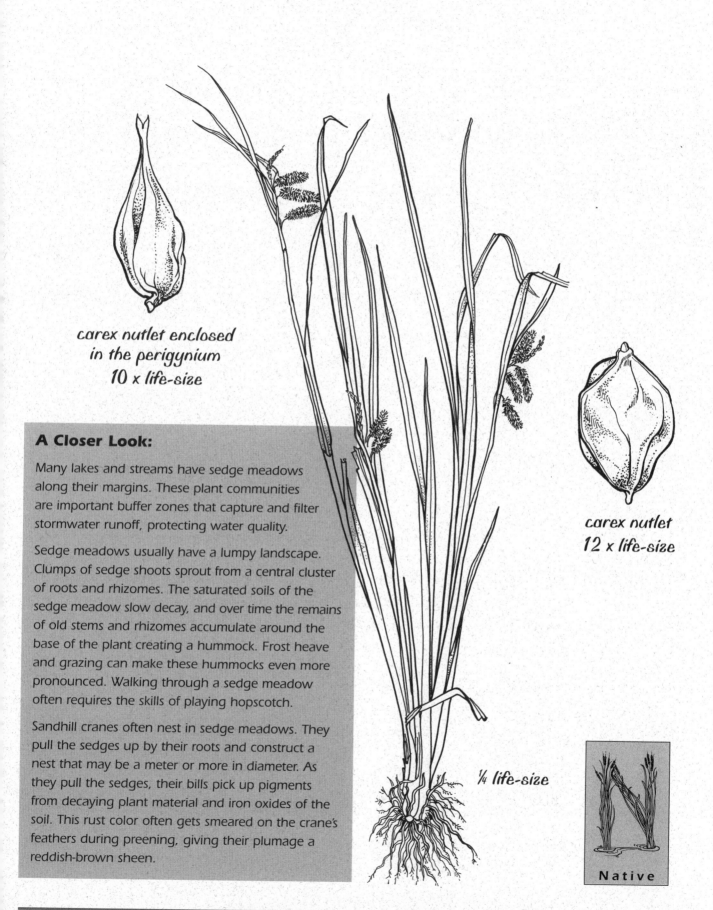

carex nutlet enclosed
in the perigynium
10 x life-size

carex nutlet
12 x life-size

¼ life-size

A Closer Look:

Many lakes and streams have sedge meadows along their margins. These plant communities are important buffer zones that capture and filter stormwater runoff, protecting water quality.

Sedge meadows usually have a lumpy landscape. Clumps of sedge shoots sprout from a central cluster of roots and rhizomes. The saturated soils of the sedge meadow slow decay, and over time the remains of old stems and rhizomes accumulate around the base of the plant creating a hummock. Frost heave and grazing can make these hummocks even more pronounced. Walking through a sedge meadow often requires the skills of playing hopscotch.

Sandhill cranes often nest in sedge meadows. They pull the sedges up by their roots and construct a nest that may be a meter or more in diameter. As they pull the sedges, their bills pick up pigments from decaying plant material and iron oxides of the soil. This rust color often gets smeared on the crane's feathers during preening, giving their plumage a reddish-brown sheen.

Native

Carex comosa (CARE-ex co-MO-sa)

Bristly sedge, bottle brush sedge

Carex – (L.) sedge; *comosa* – (L.) hairy

A canoe nosed onto shore, resting against a dense clump of sedge. Just above the robust green leaves were fruiting spikes that look like miniature bottle brushes. A close look through a hand lens revealed that the plant was bristly sedge, one of the region's most common aquatic sedges.

Description: Sturdy stems of bristly sedge emerge from the rhizome in a dense cluster. The stems range from knee-high to waist-high and support moderately wide (6-15 mm) leaf blades. The slender spike of male flowers is on a short stalk just above the female spikes. It is the female spikes that have the bottle brush appearance. This look is created by the sac-like structures around the ovaries called perigynia. Each one has a long beak with two slender teeth (1.2-2.3 mm) that curve away from each other. When these perigynia are all clustered together in a spike, the teeth create a "bristly" appearance.

Similar species: *Carex* species are notoriously difficult to identify. Mature fruit and reference to a detailed technical key are necessary.

Origin & Range: Native; widely distributed in Wisconsin; range includes eastern U.S. and parts of the west.

Habitat: Bristly sedge grows in very shallow water or moist soil. It is found in marshes, wet meadows and along the banks of lakes, ponds and streams.

Through the Year: Bristly sedge overwinters by producing new shoots on rhizomes during the fall. These remain dormant until the spring thaw. Flowering spikes are present by mid-summer and nutlets are mature by late summer. Seeds are sometimes carried to new sites by water or animals.

Value in the Aquatic Community: The nutlets of bristly sedge are eaten by a variety of waterfowl. The dense growth form makes this sedge a valuable shoreline stabilizer.

A Closer Look:

Bristly sedge is a sentinel of succession. The shoreline *Carex* species will only grow in very shallow water. As a lakeshore fills in, the sedges will start growing in areas that once had only deeper water emergents like pickerelweed or bulrush.

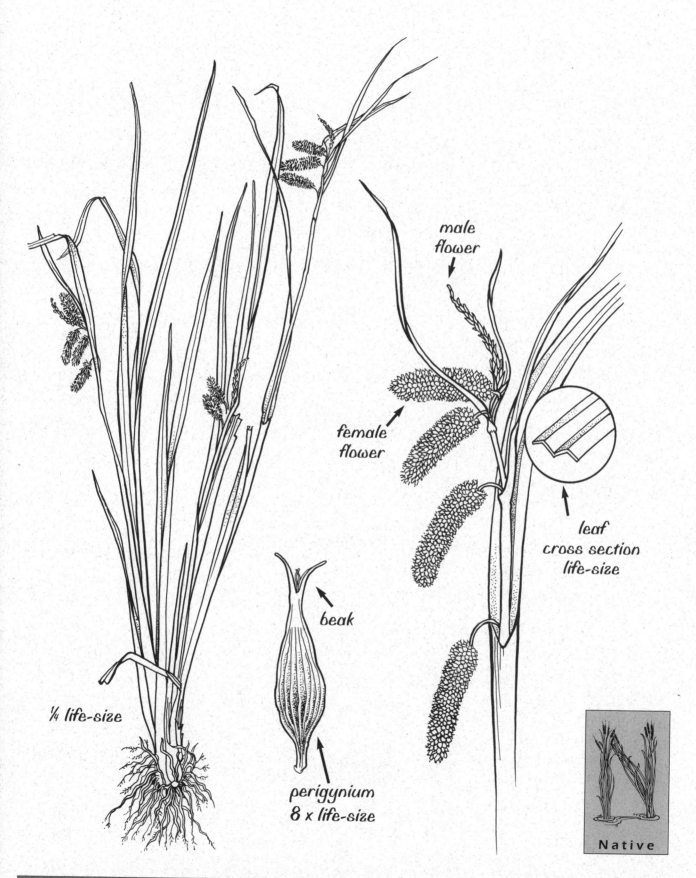

male
flower

female
flower

leaf
cross section
life-size

beak

perigynium
8 x life-size

¼ life-size

Native

EMERGENT

Dulichium arundinaceum
(du–LICH–ee–um a–run–den–ACE–ee–um)

Three-way sedge

Dulichium – name of unknown origin; *arundinaceum* – (L.) reed–like

An old man and a boy walked carefully along the shore.
The man pointed out the subtle beauty of the three-way sedge.
"When you look straight down at it, you will see that the
spiralled leaves form a perfect three-point star," he explained.

Description: The stiff stems (30 cm–1 m tall) of three-way sedge emerge from a spreading rhizome. The leaves are fairly short (5–15 cm long, 2.5–8 mm wide) and stiff, so they stand out from the plant. This gives it the appearance of a dwarf bamboo plant. Unlike the true sedges, whose stems are triangular and solid, the stem of *Dulichium* is round and hollow.

Flowers are arranged in linear spikelets (1.0–2.5 cm long) in two ranks along stalks in the upper leaf axils. The fruit is a beaked nutlet surrounded by 6–9 finely barbed bristles.

Similar species: *Dulichium arundinaceum* is the only species in the genus *Dulichium*. It is so distinctive that it is usually not confused with other plants.

Origin & Range: Native; common throughout Wisconsin; range includes most of U.S.

Habitat: Extensive beds of three-way sedge can be found in the shallow (less than 1 m) water of lakes, marshes and rivers. It grows in a variety of sediment types and can tolerate some turbidity.

Through the Year: Three-way sedge is a perennial that survives the winter as hardy rhizomes. New shoots appear in spring and flower spikes develop in leaf axils by midsummer. The brownish, beaked fruits are mature by fall.

Value in the Aquatic Community: Three-way sedge is eaten occasionally by a variety of ducks and geese. The rhizomes and shoots are also grazed by muskrats. The shallow, spreading rhizomes create interlocking stands that are effective in buffering wave action and stabilizing sediment.

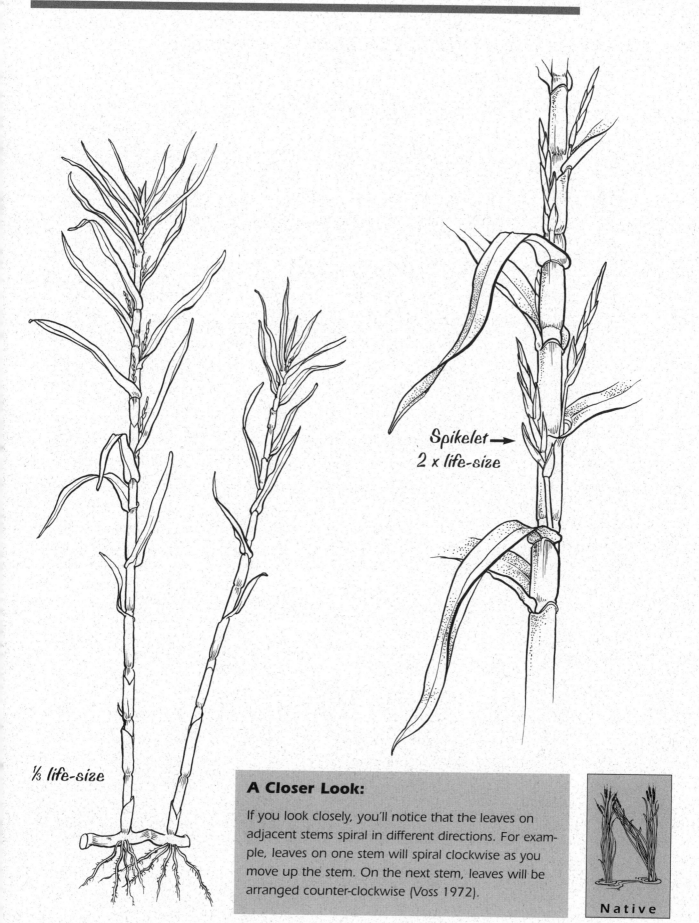

**Spikelet →
2 x life-size**

⅓ life-size

A Closer Look:

If you look closely, you'll notice that the leaves on adjacent stems spiral in different directions. For example, leaves on one stem will spiral clockwise as you move up the stem. On the next stem, leaves will be arranged counter-clockwise (Voss 1972).

Native

EMERGENT

Eleocharis acicularis (el-ee-OCK-er-res a-SIK-u-lar-us)

Needle spikerush, hairgrass

Eleocharis - (Gk.) *helos:* marsh + *charis:* grace; *acicularis* - (L.) needle-like

A puff of sand grains drifts from the siphon of a clam that has taken up residence between the dense clumps of needle spikerush. The plants cover the lake bottom, creating a carpet of grass-like turf in the shallow water.

Description: The stems of needle spikerush are slender (up to 0.25 mm thick) and rather short (3-12 cm long). They emerge in tufts from fine, spreading rhizomes. Leaves are reduced to sheaths at the base of the stem. Each stem is topped with a solitary, oval spikelet (2.5-7 mm long) that is noticeably wider than the stem.

The spikelet has a tight spiral of tiny flowers (later nutlets) covered by scales (1.5-2.2 mm long). The scales have a greenish midrib and brown-tinted margins. Mature nutlets are used for positive identification among spikerushes. It is worth the effort to look at these nutlets under magnification. The surface detail is like a fine ceramic vase and the body of the nutlet is topped with a cap called a tubercule. The nutlet of needle spikerush is rounded (0.7-1 mm long)

with a pale gray to yellow surface. The surface is textured with 8-18 lengthwise ridges and many fine crossbars. The cap rests on top, shaped like a miniature chocolate drop.

Similar species: Needle spikerush can be distinguished from other sedges by the absence of noticeable leaves and the single terminal spikelet on each stem. Needle spikerush is sometimes confused with other small spikerushes, such as **E. intermedia** or **E. olivacea.** Examining the nutlets will separate them. *Eleocharis intermedia* and *E. olivacea* also have thicker stems and the lowest scales in their spikelets are sterile.

A sterile submersed form of needle spikerush (*E. acicularis* forma *inundata*) is often found offshore. The stems become elongated and hair-like. They can be distinguished from other thread-like submersed plants by the thin rhizomes and tufted arrangement of the stems.

A Closer Look:

Needle spikerush was one of the first aquatic plants found to possess allelopathic capabilities. *Allelopathy* occurs when a species can exude a chemical that inhibits the growth of another species competing for the same space. This discovery was made when needle spikerush was grown in a greenhouse along with several species of pondweeds (Gopal and Goel 1993). Since then, several species of aquatic plants have been shown to have this capacity. The release of these inhibiting chemicals varies with environmental conditions and the species present.

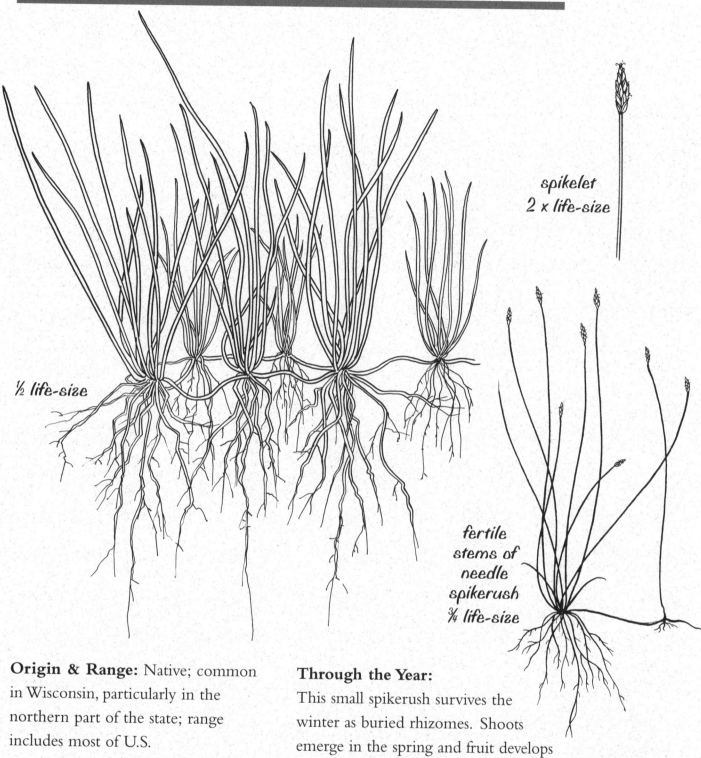

½ life-size

spikelet
2 x life-size

fertile
stems of
needle
spikerush
¾ life-size

Origin & Range: Native; common in Wisconsin, particularly in the northern part of the state; range includes most of U.S.

Habitat: Needle spikerush can be found from moist shorelines to water 2 meters deep. It is found more often on firm substrates and can tolerate some turbidity.

Through the Year:
This small spikerush survives the winter as buried rhizomes. Shoots emerge in the spring and fruit develops in the spikelets by midsummer.

Value in the Aquatic Community: Needle spikerush provides food for a wide variety of waterfowl as well as muskrats. Submersed beds offer spawning habitat and shelter for invertebrates.

Native

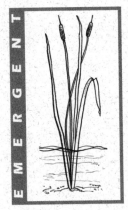

Eleocharis palustris (el-ee-OCK-er-res pa-LUS-trus)

Creeping spikerush

Eleocharis – (Gk.) *helos:* marsh + *charis:* grace; *palustris* – (L.) of marshes

A wood duck hen wove her way through a bed of spikerushes with her ducklings close behind. The brood was sheltered from breaking waves by the rushes, and boaters steered around the margin of the rush bed. The little mob fed hungrily on the high-energy food provided by the roots and rhizomes.

Description: Creeping spikerush has stems (10 cm-1 m tall) that arise singly or in small clusters from rhizomes about the same diameter as the stems (0.5-3 mm). Leaves are reduced to sheaths at the base of the stem. Each stem is topped with a single spikelet (5-25 mm long) that tapers to a point. The stem and spikelet together have the overall shape of a burning chimney match. Each spikelet has a tight spiral of flowers (later nutlets) covered by scales (2-4.5 mm). The scales are chestnut brown with pale margins. The bottom 1-3 scales are sterile (contain no flowers).

The mature nutlet is used for positive identification among spikerushes. The surface detail is remarkable under magnification and the nutlet is topped with a distinctive cap. Creeping spikerush has a golden brown nutlet (1-2 mm long) with a finely textured surface. The knob-like cap (0.4-0.7 mm) covers about half of the top of the nutlet.

Similar species: Creeping spikerush can be distinguished from other sedges by the absence of noticeable leaves and the single terminal spikelet on each stem. It can be separated from other *Eleocharis* species by examining mature fruit. A number of taxonomists now group several similar species with *E. palustris* including *E. smallii, E. calva* and *E. erythropoda.*

One aquatic spikerush is listed as **Endangered** in Wisconsin:

Square-stemmed spikerush (*Eleocharis quadrangulata*) is about the same height as creeping spikerush, but it has a thick, sharply four-angled stem.

Origin & Range: Native; common in Wisconsin; range includes northern U.S.

Habitat: Creeping spikerush is found in marshes, wet meadows, ditches and lakeshores. It grows from moist shorelines to depths of 2 meters. It is most often found on firm substrate.

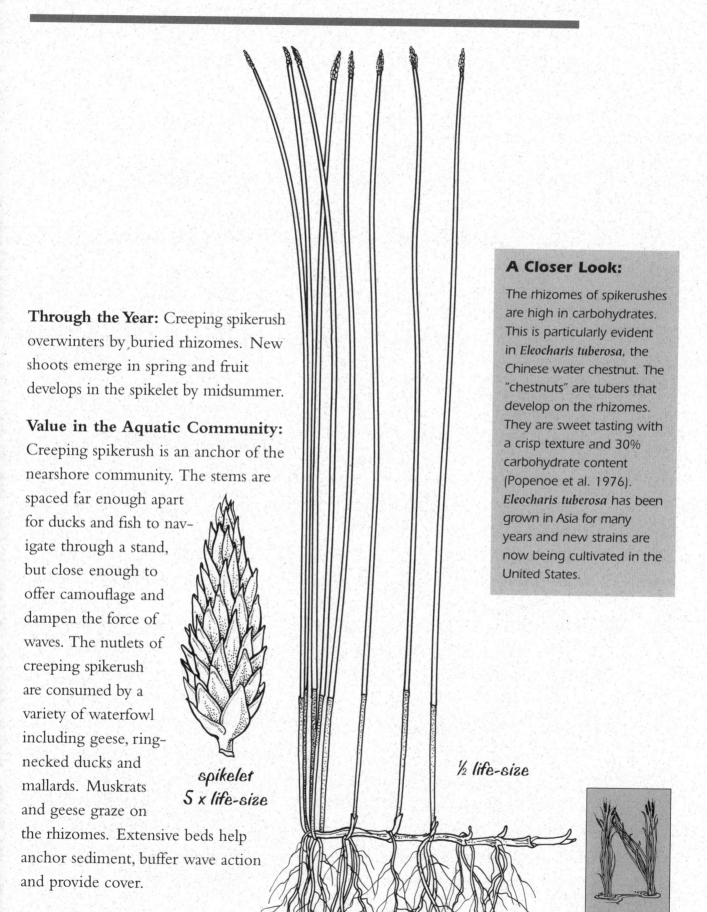

Through the Year: Creeping spikerush overwinters by buried rhizomes. New shoots emerge in spring and fruit develops in the spikelet by midsummer.

Value in the Aquatic Community: Creeping spikerush is an anchor of the nearshore community. The stems are spaced far enough apart for ducks and fish to navigate through a stand, but close enough to offer camouflage and dampen the force of waves. The nutlets of creeping spikerush are consumed by a variety of waterfowl including geese, ring-necked ducks and mallards. Muskrats and geese graze on the rhizomes. Extensive beds help anchor sediment, buffer wave action and provide cover.

spikelet
5 x life-size

½ life-size

A Closer Look:

The rhizomes of spikerushes are high in carbohydrates. This is particularly evident in *Eleocharis tuberosa*, the Chinese water chestnut. The "chestnuts" are tubers that develop on the rhizomes. They are sweet tasting with a crisp texture and 30% carbohydrate content (Popenoe et al. 1976). *Eleocharis tuberosa* has been grown in Asia for many years and new strains are now being cultivated in the United States.

N a t i v e

EMERGENT

Eleocharis robbinsii (el-ee-OCK-er-res row-BIN-see-i)

Robbins spikerush

Eleocharis – (Gk.) *helos:* marsh + *charis:* grace; *robbinsii* – named for
James W. Robbins (1801-1879)

*The stems of Robbins spikerush stuck out of the water like angular
chop sticks. Beneath the surface, fine hair-like stems trailed in the
water creating a fine network of hiding places for minnows.*

Description: Robbins spikerush has emergent stems (20-70 cm tall, 1-2 mm thick) that sprout from spreading rhizomes. The stems are stiff, slender and triangular. The leaves are reduced to a brown sheath at the base of a stem that looks as though it has been cut at an angle. The stem is tipped with a spikelet that is often difficult to see. If you squeeze the tip, nutlets will sometimes pop out. The nutlets (2-2.5 mm) are brown and urn-shaped. There is a flattened knob on top of the nutlet that is about half as long as the nutlet itself. Underwater stems are often produced that are as fine as hair. These stems are sterile and float in the water from the base of the plant.

Similar species: Robbins spikerush can be distinguished from other sedges by the absence of noticeable leaves and the single terminal spikelet on each stem. It can be separated from other *Eleocharis* species by examining mature fruit.

Origin & Range: Native; scattered locations, primarily in northern Wisconsin; range includes eastern U.S. Robbins spikerush is listed as a **Special Concern** species in Wisconsin.

Habitat: Robbins spikerush grows from moist shorelines to water over a meter deep. It is usually found in soft-water, low pH lakes.

Through the Year: Robbins spikerush survives the winter with hardy rhizomes. New shoots emerge as the water warms in spring. Reproduction from seed may occur under favorable conditions. Spikelets form on emergent stems by midseason. Mature nutlets develop by mid- to late summer.

Value in the Aquatic Community: Stems, rhizomes and nutlets are consumed by a variety of waterfowl. Muskrats also graze on stems and rhizomes. The fine submersed stems offer habitat for invertebrates and small fish.

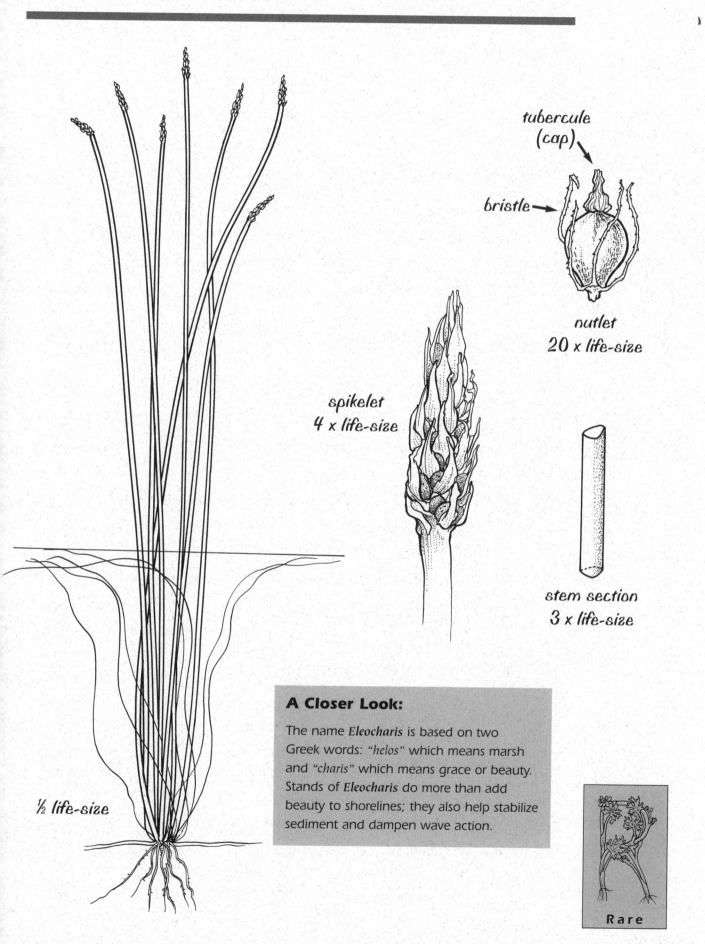

tubercule
(cap)

bristle

nutlet
20 x life-size

spikelet
4 x life-size

stem section
3 x life-size

½ life-size

A Closer Look:

The name *Eleocharis* is based on two
Greek words: *"helos"* which means marsh
and *"charis"* which means grace or beauty.
Stands of *Eleocharis* do more than add
beauty to shorelines; they also help stabilize
sediment and dampen wave action.

Rare

EMERGENT

Equisetum fluviatile
(eh-kwa-SEE-tum flew-vee-AH-till-ee)

Water horsetail, pewterwort, joint rush

Equisetum – (L.) *equus*: horse + *seta*: bristle; *fluviatile* – (L.) *fluvius*: river

The ridged stem of the water horsetail felt as gritty as a nail file. With a firm tug, the jointed stem popped apart. The segmented stem kept the plant from being uprooted. Who would have guessed that 200 million years ago dinosaurs grazed on the ancestors of these plants?

Description: Water horsetail has jointed stems that emerge from a buried rhizome. The hollow stems (up to 1 m or more tall) are ridged and stiff, with a high silica content. They have a central cavity that takes up about 80% of the stem diameter. Each stem is jointed and can be pulled apart at the nodes which makes a satisfying snap, like popping bubble wrap. The leaves are reduced to a membrane with teeth that encircles each node. The 15-20 teeth (1.5-3.0 mm long) are pointed and dark brown to black with a pale margin. Branching may occur at the nodes depending on growing conditions. Sterile and fertile stems look alike, except fertile stems produce a terminal cone. The oval cone produces spores that germinate on moist soil.

Similar species: Horsetails are easy to distinguish from sedges and rushes by their jointed stems. Water horsetail can be separated from other horsetail species by its large central cavity (with no side cavities) and the appearance of the teeth.

Origin & Range: Native; common in Wisconsin; range includes northern U. S.

Habitat: Water horsetail is found in marshes and shallow water of lakes and ponds. It usually occurs in water less than 1 meter deep.

Through the Year: Winter survival is dependent on the buried rhizomes. New shoots are produced in spring. Fertile stalks are evident by early summer. Stems die back in fall after a hard frost.

Value in the Aquatic Community: Water horsetail provides food for waterfowl (primarily geese) and is grazed by ruffed grouse and moose. Recent research has shown water horsetail is a primary food source for trumpeter swans on their breeding grounds in Alaska. It dominates the post-hatch diet of both adults and young. The continuous development of new shoots offers a reliable and easily accessible source of food for the cygnets (Grant et al. 1994).

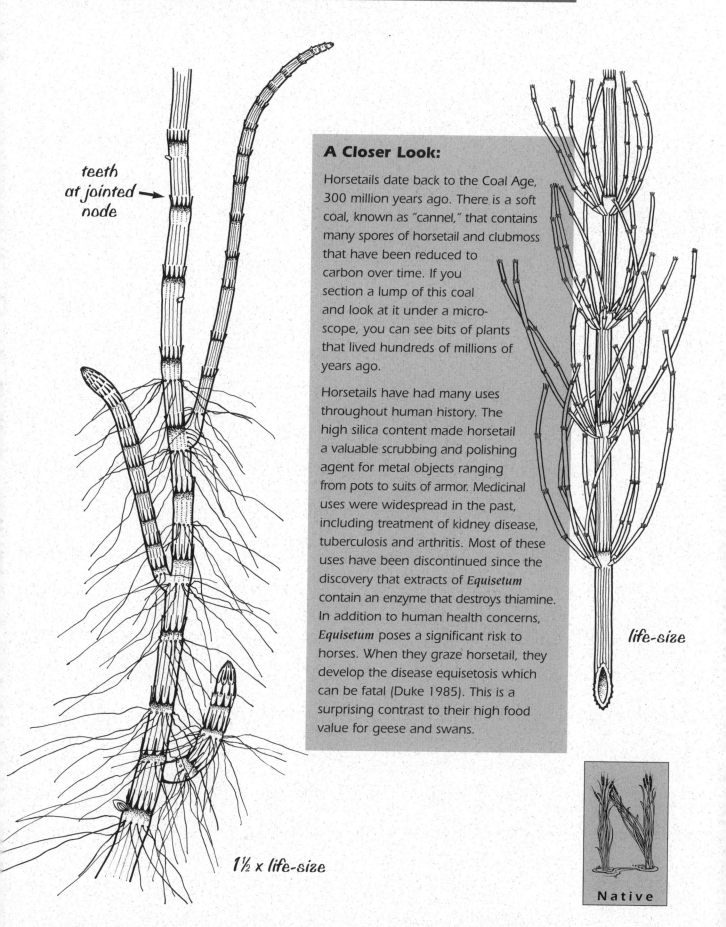

teeth
at jointed →
node

A Closer Look:

Horsetails date back to the Coal Age, 300 million years ago. There is a soft coal, known as "cannel," that contains many spores of horsetail and clubmoss that have been reduced to carbon over time. If you section a lump of this coal and look at it under a microscope, you can see bits of plants that lived hundreds of millions of years ago.

Horsetails have had many uses throughout human history. The high silica content made horsetail a valuable scrubbing and polishing agent for metal objects ranging from pots to suits of armor. Medicinal uses were widespread in the past, including treatment of kidney disease, tuberculosis and arthritis. Most of these uses have been discontinued since the discovery that extracts of *Equisetum* contain an enzyme that destroys thiamine. In addition to human health concerns, *Equisetum* poses a significant risk to horses. When they graze horsetail, they develop the disease equisetosis which can be fatal (Duke 1985). This is a surprising contrast to their high food value for geese and swans.

life-size

1½ x *life-size*

Native

E
M
E
R
G
E
N
T

Glyceria borealis (gly-CER-ee-a bor-ee-AL-is)

Northern manna grass

Glyceria – (Gk.) *glykeros*: sweet (referring to taste of the grain);
borealis – (L.) northern, from *Boreas*, Greek god of the north wind

Four gadwalls meandered through the beds of northern manna grass. Their attention was focused on discovering and scooping up the protein-rich grain the plant dispensed.

Description: Grasses have a unique structure. Once you become familiar with the various parts, they are not difficult to identify. The basic unit of structure is a spikelet composed of one or more flowers called florets. At the base of each spikelet are two bracts known as glumes. Within the spikelet each floret has two scales enclosing the reproductive parts. The appearance of these scales is often a primary feature in separating different grasses. The outer scale is called the lemma and the inner scale is called the palea.

Northern manna grass often grows in clumps that expand in a sprawling manner, rooting at the nodes. The stems can be up to 1 meter or more high and have leaves 2–5 mm wide. The leaf sheath is closed rather than open, an unusual feature among grass species. Northern manna grass can also produce slender, floating leaves when it is growing in the water. These leaves are limp and have a water-resistant upper surface.

The overall flower spike is slender, with

Grass spikelet
4 x life-size

florets

palea

Grass floret
8 x life-size

lemma

several spikelets on each 8–12 cm long branch. The individual linear spikelets are 1–2 cm long. The lemmas (3–4 mm long) are rounded with a thin, brittle margin and tip. The surface of the lemma is shiny and smooth between the raised nerves, giving it a ridged or corrugated appearance.

Similar species: Manna grass can be distinguished from other grasses by the combination of a closed leaf sheath and strongly ridged lemmas. Northern manna grass can be separated from other manna grasses by the size and shape of the spikelet and the surface features of the lemma. Floating leaves may be confused with floating leaves of wild rice. Manna grass has fine hairs on the upper leaf surface, while wild rice is smooth.

A Closer Look:

Reed sweet-grass (*Glyceria maxima*), a relative of *G. borealis*, is being tested in Europe as a wastewater treatment plant. Constructed wetlands planted with *G. maxima* have been effective in nutrient absorption and the harvested *Glyceria* has been used as animal fodder.

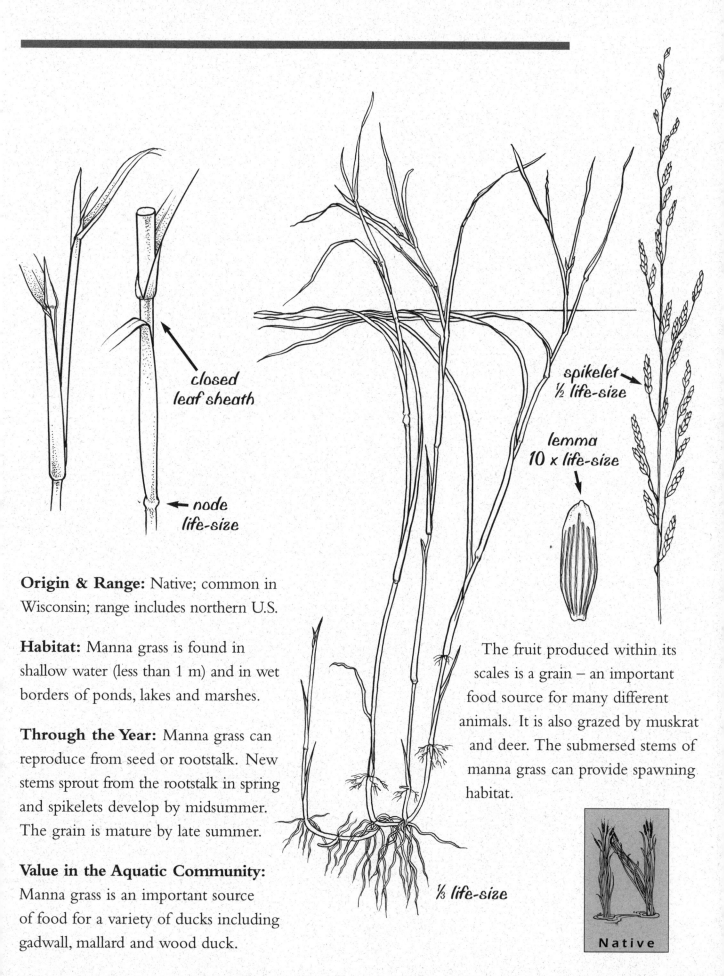

closed
leaf sheath

← node
life-size

spikelet
½ life-size

lemma
10 x life-size

Origin & Range: Native; common in Wisconsin; range includes northern U.S.

Habitat: Manna grass is found in shallow water (less than 1 m) and in wet borders of ponds, lakes and marshes.

Through the Year: Manna grass can reproduce from seed or rootstalk. New stems sprout from the rootstalk in spring and spikelets develop by midsummer. The grain is mature by late summer.

Value in the Aquatic Community: Manna grass is an important source of food for a variety of ducks including gadwall, mallard and wood duck.

The fruit produced within its scales is a grain – an important food source for many different animals. It is also grazed by muskrat and deer. The submersed stems of manna grass can provide spawning habitat.

⅓ life-size

Native

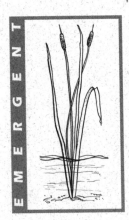

Iris versicolor (EYE-ris VER-si-col-or)

Northern blue flag

Iris – (Gk.) *Iris*, Goddess of the Rainbow; *versicolor* – (L.) varied color

Blue flag is one of the aristocrats of the plant world. From across the lake, the brilliant blue commands attention. Its prominent bloom is a proud emblem for its namesake, Iris, the Greek Goddess of Rainbows.

Description: The leaves and flower stalks of northern blue flag emerge from a stout rhizome that is very shallow and sometimes exposed above the sediment. The sword-like leaves (0.5-5 cm wide) grow in flattened, fan-like clusters. The flower stalks (20-80 cm high) are taller than the leaves, creating the "flag" appearance.

The flowers (6-8 cm wide) range from indigo blue to lavender. The three outer, petal-like lobes of the blossom are sepals. They may be entirely blue or have a small, greenish-yellow spot at the base of the blade. The stalks that connect stigmas to ovaries, called styles, are also flattened and petal-like and arch over the sepals. The three upright petals are shorter than either the sepals or styles, but similar in color.

Similar species: Southern blue flag (*Iris virginica* var. *shrevei*) is similar in appearance and has an overlapping range with northern blue flag. The flowers of southern blue flag have a hairy, bright yellow spot at the base of the sepals, the petals are longer than the styles, and the seeds are dull with an irregularly pitted or unpitted surface. When blue flag is not in flower, the leaves could be confused with sweetflag (*Acorus calamus*). However, sweetflag's leaves have a wavy margin and an off-center midrib.

Origin & Range: Native; common, especially in northern Wisconsin; range includes eastern to central U.S.

Habitat: Northern blue flag is found in wetlands, stream banks and shallow water (less than 1 m deep) of ponds and lakes.

Through the Year: Northern blue flag is a perennial that can overwinter by hardy rhizomes. During the first year or two, rhizomes produce only leaves. After the rhizome is well established, flower stalks are produced next to leaf clusters. The flowers bloom from May through July. Seeds develop in the capsular fruit by late summer and can be found floating on the water during the fall.

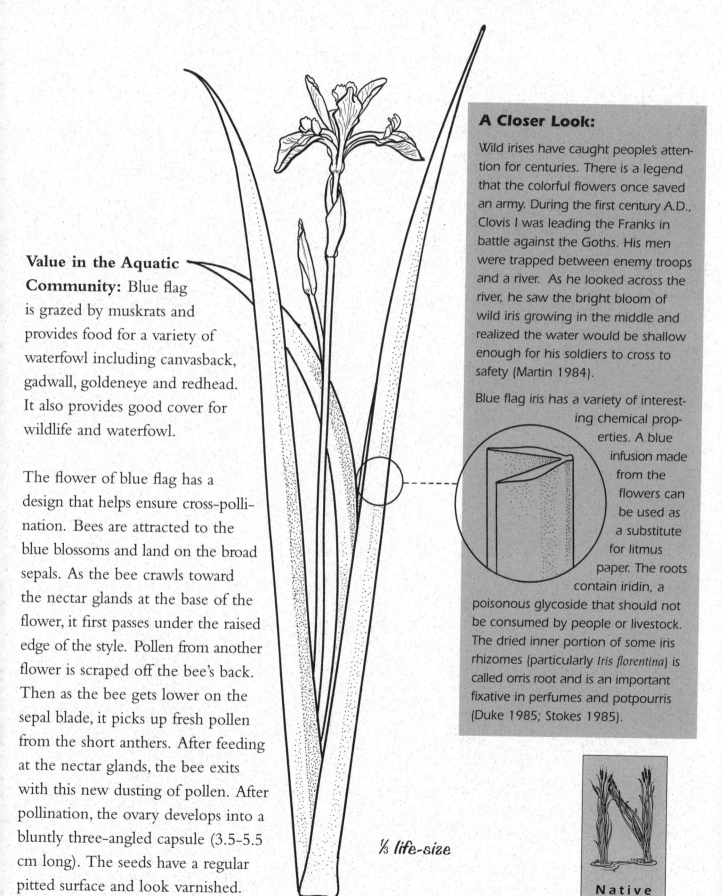

Value in the Aquatic Community: Blue flag is grazed by muskrats and provides food for a variety of waterfowl including canvasback, gadwall, goldeneye and redhead. It also provides good cover for wildlife and waterfowl.

The flower of blue flag has a design that helps ensure cross-pollination. Bees are attracted to the blue blossoms and land on the broad sepals. As the bee crawls toward the nectar glands at the base of the flower, it first passes under the raised edge of the style. Pollen from another flower is scraped off the bee's back. Then as the bee gets lower on the sepal blade, it picks up fresh pollen from the short anthers. After feeding at the nectar glands, the bee exits with this new dusting of pollen. After pollination, the ovary develops into a bluntly three-angled capsule (3.5–5.5 cm long). The seeds have a regular pitted surface and look varnished.

⅓ *life-size*

A Closer Look:

Wild irises have caught people's attention for centuries. There is a legend that the colorful flowers once saved an army. During the first century A.D., Clovis I was leading the Franks in battle against the Goths. His men were trapped between enemy troops and a river. As he looked across the river, he saw the bright bloom of wild iris growing in the middle and realized the water would be shallow enough for his soldiers to cross to safety (Martin 1984).

Blue flag iris has a variety of interesting chemical properties. A blue infusion made from the flowers can be used as a substitute for litmus paper. The roots contain iridin, a poisonous glycoside that should not be consumed by people or livestock. The dried inner portion of some iris rhizomes (particularly *Iris florentina*) is called orris root and is an important fixative in perfumes and potpourris (Duke 1985; Stokes 1985).

Native

Juncus effusus (JUNK-us ef-FEW-sus)

Soft rush, common rush, mat rush

Juncus – (L.) rush; *effusus* – (L.) overflowing

A swamp sparrow perched on a clump of soft rush, picking at the fruit. The stalk, with its tuft of tiny lily-like flowers, bowed under the bird's weight.

Description: Soft rush has smooth, cylindrical stems (1 m or more tall) that emerge from a dense rootstalk. The leaves are reduced to reddish-brown sheaths at the base of the stem. Although the flower clusters appear to grow from the side of each stem, the "stem" portion beyond the flowers is actually a slender floral leaf (15-25 cm long). Each flower is poised on the end of a slender stalk (4-10 cm long). The six straw-colored tepals (three sepals and three petals) eventually surround a capsular fruit.

Similar Species: Rushes are sometimes mistaken for grasses or sedges. However, the flowers appear early in the season and a close look at them will quickly distinguish a rush from a grass.

Origin & Range: Native; common throughout Wisconsin; range includes most of the U.S.

Habitat: Soft rush is often found growing in large clumps in wet meadows and the shallow water of lake shores.

Through the Year: New stems emerge from the overwintering rootstalk in spring. Flowers are produced by early summer. The sepals and petals persist and eventually surround a seed-filled capsule.

Value in the Aquatic Community: Clumps of soft rush provide cover and seeds for waterfowl, game birds, marsh birds and song birds. New shoots are grazed by muskrat. The base of the stems offers spawning habitat, particularly for rock bass.

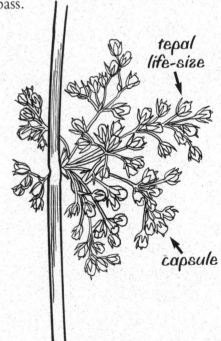

tepal life-size

capsule

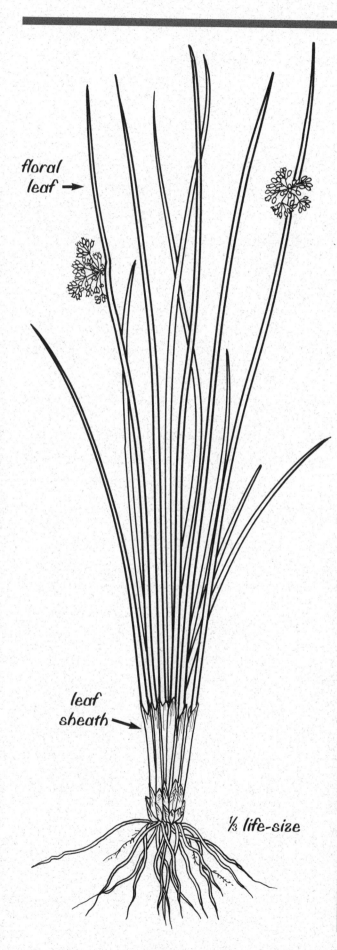

floral
leaf →

leaf
sheath →

⅓ life-size

A Closer Look:

Soft rush is also known as mat rush. It has been a valuable weaving material in Asia for many years. In Japan, Taiwan, Korea and China, it is cultivated as a wetland crop and harvested for use in binding, basketry and floor matting. Soft rush has also had an important historic use in "rushlights" for over 2,000 years. The outer skin was peeled off most of the stem, revealing the gauzy pith. This was dipped in tallow, creating a slender candle with an even burning rate. Rushlights made a comeback in England during World War II. Candles were in short supply during the blackouts, so soft rush was gathered from wetlands and made into rushlights (Mabey 1977).

Native

EMERGENT

Juncus pelocarpus (JUNK-us pel-o-CARP-us)

Brown-fruited rush

Juncus - (L.) rush; *pelocarpus* - (Gk.) *pelos:* mud + *carpus:* fruit

A bluegill hovers in the shallow water, nestled in the soft turf of short, submersed rushes. It holds its ground, guarding the sandy nest site. The rushes surrounding the small depression offer a buffer from the waves and current. They will soon provide cover for the hatchlings.

Description: The submersed form of brown-fruited rush (*Juncus pelocarpus* f. *submersus*) has tufts of tapered green leaves (<3 mm wide), connected by fine horizontal stems. The outer leaves are cupped around the inner ones at their base. If a leaf is held up to the light, fine cross lines can be seen with a hand lens. The submersed form does not produce flowers and simply spreads by rhizomes.

The emergent form of brown-fruited rush usually grows on the shoreline rather than in the water. The leaf blades are rounded with regularly spaced cross-partitions. These can be seen by holding a leaf up to the light or felt by running a fingernail along the leaf surface.

Numerous pale brown flowers are borne singly (occasionally in twos or threes) along one side of spreading branches. The branched flower stalks (5-15 cm long) make up about 25% or more of the overall height of the plant (10-30 cm). Each flower has three sepals and three petals that are scale-

like with a chaffy margin. There are six stamen, one behind each sepal and petal. The fruit is a capsule (2.4-3.1 mm) that tapers to a slender beak.

A distinctive feature of brown-fruited rush is the presence of awl-shaped bulblets, which often take the place of some or all of the flowers.

Similar species: The submersed form of brown-fruited rush could be confused with other "turf" forming submersed plants such as creeping spearwort (*Ranunculus flammula*) or needle spikerush (*Eleocharis acicularis*). Creeping spearwort has arched, above-ground rhizomes called stolons that are the same width as the leaves. Needle spikerush has fine rhizomes, but the "leaves" (actually stems) are not cupped around each other like those of brown-fruited rush. The emergent form of brown-fruited rush can be distinguished from other rushes by the combination of cross-partitioned leaves and presence of bulblets.

A Closer Look:

The bulblets that often replace the flowers of brown-fruited rush are "viviparous." This term is often associated with snakes that bear live young. In the case of brown-fruited rush, it means the bulblet can sprout while still attached to the parent plant.

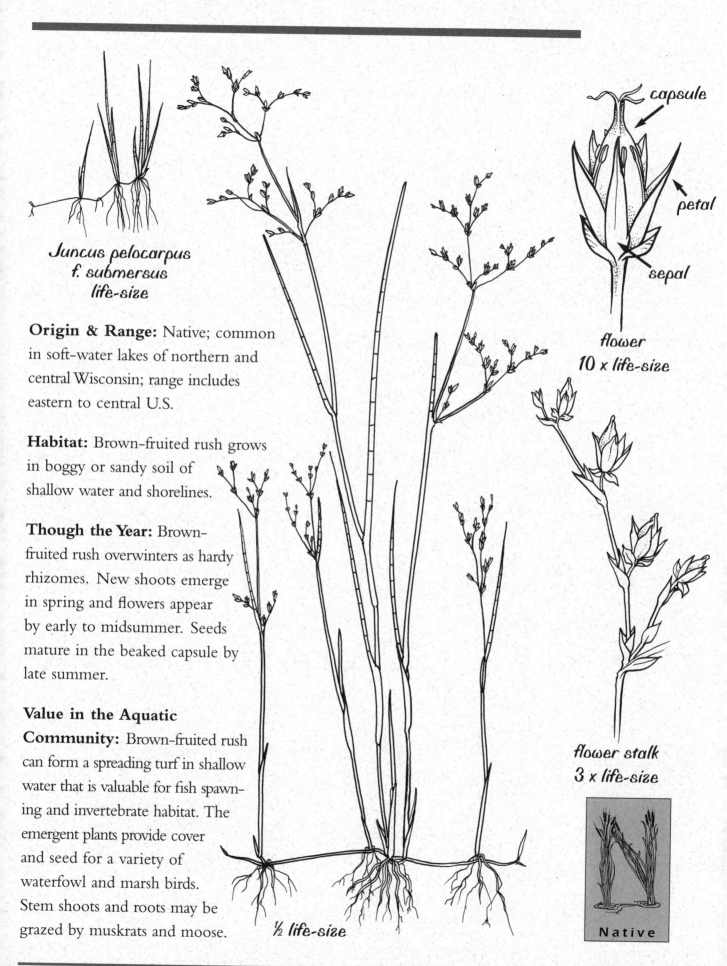

capsule

petal

sepal

flower
10 x life-size

Juncus pelocarpus
f. submersus
life-size

Origin & Range: Native; common in soft-water lakes of northern and central Wisconsin; range includes eastern to central U.S.

Habitat: Brown-fruited rush grows in boggy or sandy soil of shallow water and shorelines.

Though the Year: Brown-fruited rush overwinters as hardy rhizomes. New shoots emerge in spring and flowers appear by early to midsummer. Seeds mature in the beaked capsule by late summer.

Value in the Aquatic Community: Brown-fruited rush can form a spreading turf in shallow water that is valuable for fish spawning and invertebrate habitat. The emergent plants provide cover and seed for a variety of waterfowl and marsh birds. Stem shoots and roots may be grazed by muskrats and moose.

flower stalk
3 x life-size

Native

½ life-size

EMERGENT

Leersia oryzoides (LEER-zee-a or-e-ZOID-eez)

Rice cut-grass

Leersia – named for Johann D. Leer, a German botanist (1727-1774);

oryzoides – (L.) *oryza:* rice + *oides:* resembling

The disturbed shoreline was a prime spot for rice cut-grass to grow. Those who ventured into the stand quickly realized the reason it was called "cut" grass. They became acquainted with its sharp edges and spines, and later the art of peeling its velcro-like seeds from their clothing and pet's fur.

A Closer Look:

Rice cut-grass "volunteers" in disturbed areas and crowds out plants which may have been seeded intentionally. It has a competitive advantage because it is one of the first grasses to start growing in spring.

Description: The sprawling stems of rice cut-grass emerge from a slender rootstalk and grow from knee- to shoulder-high, depending on soil and moisture conditions. The leaf blades (15-30 cm long, 6-15 mm wide) have very rough edges with stiff spines that can tear skin or clothing.

The branches bearing spikelets (see discussion of grass structures under *Glyceria*) are slender and spreading. The spikelets are in clusters of 3-8 on the ends of the branches. Each spikelet (4-5.5 mm long, 1.4-1.8 mm wide) is flattened and overlaps about half of the one above it (like tiles on a roof).

The surface of the lemna has stiff hairs that tightly adhere to clothing or fur.

lemma

8 x life-size

Spikelets on the open branches are not always fertile. Others form inside the upper leaf sheath and release their seeds as the stems decay at the end of the growing season.

Similar Species: There are two less frequent *Leersia* species that occur in this region. **White grass** (*Leersia virginica*) usually grows in partial shade and doesn't have stiff hairs on the leaf margins. **Catchfly grass** (*Leersia lenticularis*) has a stout rootstalk and spikelets (3-4 mm wide) that are twice as wide as those of rice cut-grass. The leaves of catchfly grass are smooth and don't have stiff hairs and rough edges.

Origin & Range: Native; common throughout Wisconsin; range includes most of U.S.

Habitat: Rice cut-grass grows in marshes, wet meadows and on the borders of lakes, ponds and streams.

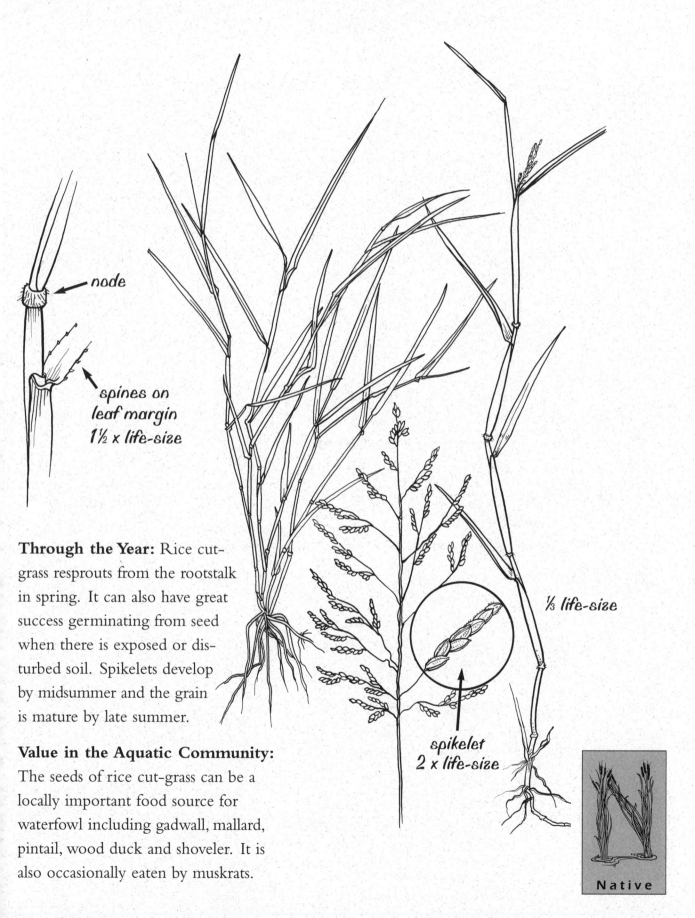

node

spines on leaf margin 1½ x life-size

Through the Year: Rice cut-grass resprouts from the rootstalk in spring. It can also have great success germinating from seed when there is exposed or disturbed soil. Spikelets develop by midsummer and the grain is mature by late summer.

Value in the Aquatic Community: The seeds of rice cut-grass can be a locally important food source for waterfowl including gadwall, mallard, pintail, wood duck and shoveler. It is also occasionally eaten by muskrats.

⅓ life-size

spikelet 2 x life-size

Native

EMERGENT

Phalaris arundinacea (FA-lar-is a-run-din-ACE-ee-a)

Reed canary grass

Phalaris – (L.) *phalaros:* having a white spot; *arundinacea* – (L.) reedlike

The edge of the lake had seen its share of changes. The forest had been replaced by fields and new houses were springing up. Thick stands of reed canary grass were well established there, capitalizing on the alterations.

Description: Dense colonies of reed canary grass are formed by spreading rhizomes. The stems are stout and can be over 1 meter tall. The grayish-green leaf blades (10-20 cm long, 10-15 mm wide) are flat and make a good "leaf whistle." The spikelets (see discussion of grass structures under *Glyceria*) form dense clusters on stalks above the leaves. Each spikelet is one-flowered. The spikelets are initially green and are sometimes tinged with purple. They become beige-colored later in the growing season.

Similar Species: Reed canary grass is sometimes confused with **bluejoint grass** (*Calamagrostis canadensis*). However, the spikelets of bluejoint are in a loose, open arrangement and the nodes have a bluish to purple cast.

Origin & Range: Both native and introduced strains of reed canary grass were present at one time, but little evidence of the native strain remains; common throughout Wisconsin; range includes northern and western portions of U.S.

Habitat: Reed canary grass is found on lakeshores, streambanks, marshes and exposed moist ground. Its success is favored by disturbance. Although it is usually a shoreline plant, reed canary grass can survive in knee-deep water by sprouting "water roots" on the submersed portion of the stem.

Through the Year: Reed canary grass is one of the first grasses to sprout in spring. It overwinters by hardy rhizomes. The spikelets are formed by midsummer and mature by late summer.

Value in the Aquatic Community: Reed canary grass is an effective shoreline stabilizer. Although it has low food value it does offer summer cover and habitat for waterfowl at disturbed sites. However, its tendency to mat down in winter means it provides little winter cover for wildlife. The nutrient level of reed canary grass is average, but its poor digestibility and presence of alkaloids limit its food value for wildlife. A close relative of reed canary grass is canary grass (*Phalaris canariensis*) which is a cultivated annual that yields seeds that are a main ingredient in commercial bird seed (Voss 1972).

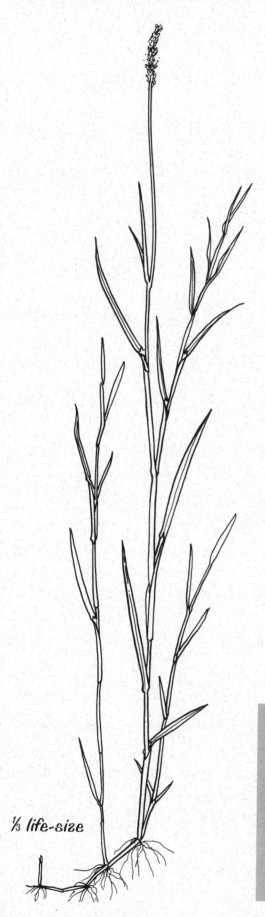

⅓ life-size

ligule

leaf blade

leaf sheath
1½ x life-size

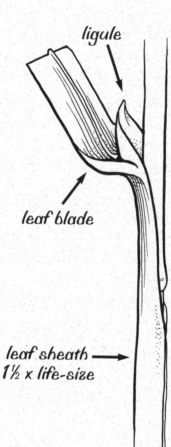

clusters
of
spikelets
life-size

A Closer Look:

Reed canary grass is the "poster plant" for disturbed wet sites. It forms dense stands in places where clearing, grading, siltation, filling or other disruptions have created an opening with moist soil. A Eurasian strain has been used for erosion control and lowland pastures, but its aggressive growth has made it a threat to native wetland plants.

Native

Phragmites australis (frag-MIDE-eez aus-TRAL-es)

(formerly known as Phragmites communis)

Giant reed, common reed

Phragmites – (Gk.) growing in hedges; *australis* – (L.) southern

*Giant reeds towered over a boat idling along the shore.
The feather-like plumes at the top of the plant swayed in
the warm summer breeze. The dense thicket of the great
plants absorbed the hum of the outboard motor.*

Description: Giant reed has stems (2-4 m tall) that grow out of stout rhizomes. The leaves (up to 60 cm long and 2-3 cm wide) may wave in the wind, like pennants. Some of the stems are topped with spreading clusters of spikelets (see discussion of grass structure under *Glyceria*). Each spikelet (10-15 mm long) has 3-7 florets and long silky hairs that give the overall flowering portion a feather duster appearance.

Similar Species: Giant reed is sometimes confused with reed canary grass because they both form dense stands at disturbed wetland sites. However, they are easily distinguished by a number of features: reed canary leaves are shorter (10-20 cm long) and narrower (10-15 mm wide), the spikelets of reed canary have only one flower, and reed canary is shorter in overall height (1-1.5 m).

Origin & Range: Native; scattered locations in Wisconsin; range includes most of U.S.

Habitat: Giant reed grows on wet shores and disturbed sites in a variety of wetlands. It can grow in water up to 2 meters deep on a variety of sediments.

Through the Year: Giant reed is a perennial that overwinters by forming buds on the rhizome. Buds are formed during the summer months and are dormant until the following spring. Stems shoot up quickly when the soil warms up and spikelets form by mid-summer. Long white hairs give the spikelets a feathery appearance by late summer. Fertile grains are not always produced and giant reed expands primarily from the rhizomes.

Value in the Aquatic Community: The rhizomes of giant reed help stabilize shorelines, but the dense stands can exclude other beneficial plants. Giant reed provides little food for waterfowl, but is grazed by muskrats. The standing winter stalks offer some cover for wildlife.

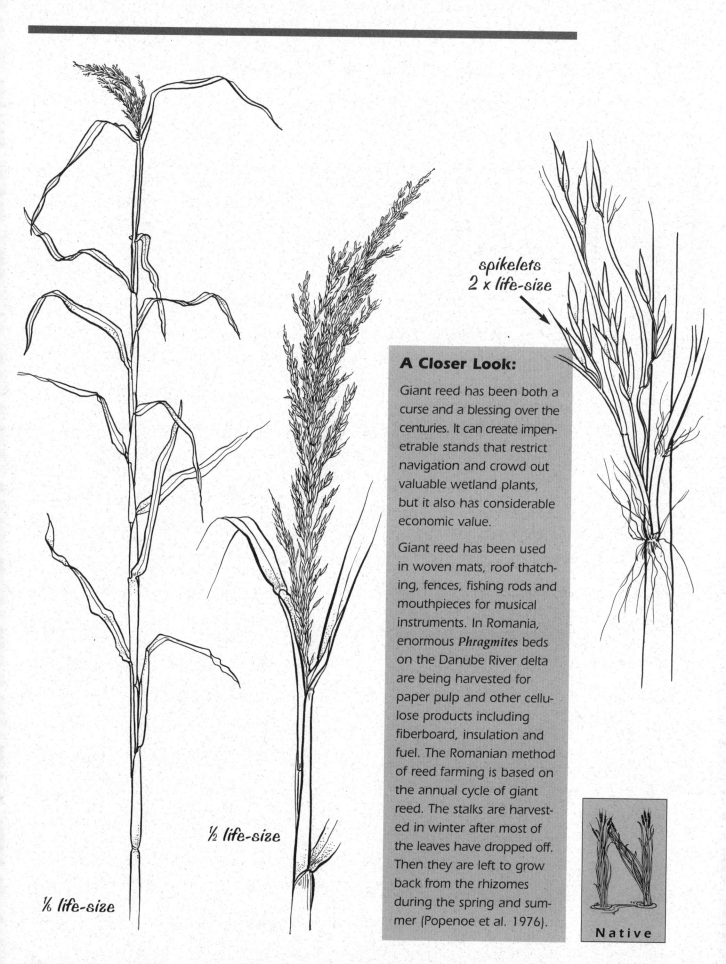

spikelets
2 x life-size

A Closer Look:

Giant reed has been both a curse and a blessing over the centuries. It can create impenetrable stands that restrict navigation and crowd out valuable wetland plants, but it also has considerable economic value.

Giant reed has been used in woven mats, roof thatching, fences, fishing rods and mouthpieces for musical instruments. In Romania, enormous *Phragmites* beds on the Danube River delta are being harvested for paper pulp and other cellulose products including fiberboard, insulation and fuel. The Romanian method of reed farming is based on the annual cycle of giant reed. The stalks are harvested in winter after most of the leaves have dropped off. Then they are left to grow back from the rhizomes during the spring and summer (Popenoe et al. 1976).

½ *life-size*

⅛ *life-size*

Native

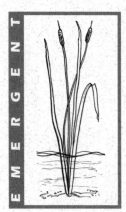

EMERGENT

Scirpus acutus (SKER-pus a-CUE-tus)

Hardstem bulrush

Scirpus – (L.) bulrush; *acutus* – sharp

Bulrushes stretched the length of the bay. A canoe maneuvered through the edge of the bed, and a loon slipped out the other end. The dense stand of bulrush dampened the waves and provided a haven from probing eyes.

Description: Hardstem bulrush has tall, sturdy stems (1–3 m tall, 0.5–1 cm wide) that emerge from a shallow rhizome. The cylindrical, olive-green stems are firm when pressed between your fingers. This firmness is due to many small chambers that fill the stem.

The stems appear to be leafless, but leaf sheaths and sometimes short blades are present near the base of each stem. There is also a floral leaf called a bract that looks like a continuation of the tip of the stem. The spikelets emerge in the angle of this bract and the end of the stem.

The oval spikelets (about 2–5 times as long as wide) are clustered on the ends of rather stiff stalks. The scales (3–4 mm long) of the spikelet are spirally arranged and have distinctive surface features. Each scale is a dull, grayish-brown with shiny red flecks and a fringed margin. Nutlets (2.2–2.5 mm long) develop under the scales as the growing season progresses.

Similar Species: Hardstem bulrush is part of the sedge family. It can be separated from grasses and rushes by looking at the spikelets, which have a spiral arrangement of scale-covered flowers and nutlets. Hardstem bulrush is most similar in appearance to softstem bulrush (*Scirpus validus*). However, softstem has a bluish-green, "squeezable" stem filled with larger chambers. The spikelets of softstem are on more relaxed stalks and the scales are a warm reddish-brown. The surface of the softstem scales is also different. They are shiny, often with a green midrib, and any red flecks are restricted to the midrib area.

Origin & Range: Native; distributed primarily in northern and eastern Wisconsin; range includes most of the U.S.

Habitat: Hardstem bulrush can be found in wetlands, lakes, ponds and streams. It is usually growing in water less than 2 meters deep, but it has been found considerably deeper. Hardstem shows a preference for firm substrate with good water movement in the root zone (Eggers and Reed 1987).

Through the Year: Hardstem bulrush overwinters by hardy rhizomes and dormant winter shoots. Large colonies may develop through rhizome expansion. New shoots sprout in spring and spikelets are formed by midsummer. Nutlets mature by late summer. Some reproduction from seed occurs when exposed mud flats are available for seed germination and seedling development.

Value in the Aquatic Community: Hardstem bulrush offers habitat for invertebrates and shelter for young fish, especially northern pike. The nutlets are consumed by a wide variety of waterfowl, marsh birds (including bitterns, herons, rails) and upland birds. Stems and rhizomes are eaten by geese and muskrats. Bulrushes also provide nesting material and cover for waterfowl, marsh birds and muskrats.

⅓ life-size

A Closer Look:

Over the centuries, bulrush has remained one of the most popular plants for weaving mats and baskets. The long, leafless stems are leathery and pliable. They are also straight and don't have joints, which makes weaving easier. When securely woven, a watertight container can be produced. Bulrushes are in prime condition to be harvested for weaving when the spikelets are in bloom (Mabey 1977).

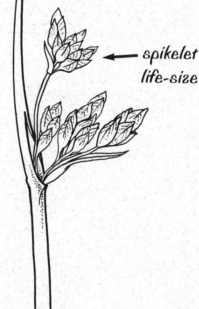

floral leaf (bract)

spikelet life-size

Native

EMERGENT

Scirpus americanus (SKER-pus a-mer-e-KAN-us)

Three-square, chairmaker's rush

Scirpus – (L.) bulrush; *americanus* – American

Snow geese returned to the lake every year to feed on the beds of three-square. These triangular-stemmed plants had always grown in the same spot. In years past, the rush rhizomes and seeds provided sustenance to the ancestors of this flock as they traveled south.

Description: Three-square has moderately tall (up to 1.5 m), sharply triangular stems that emerge from a firm rhizome. Short, inconspicuous leaves sheath the base of each stem. There is also a pointed bract (4–15 cm long) that looks like a continuation of the end of the stem. The spikelets emerge in a tight cluster in the angle of this bract and the end of the stem.

Each oval spikelet (6–12 mm long) has a spiral of reddish-brown, sharp-tipped scales. Nutlets (2.5–3 mm long) develop under the scales. Each nutlet has a stubby beak and bristles about two-thirds as long as the nutlet.

Similar Species: Three-square has two close relatives that look similar.

Torrey's three-square (*Scirpus torreyi*) grows in similar habitats, but can be distinguished by the spikelets with yellowish-brown scales, nutlets with longer beaks, soft rhizomes and blunt-tipped bracts on the end of the stems. Torrey's three-square is listed as a **Special Concern** species in Wisconsin.

Olney's three-square (*Scirpus olneyi*) grows in salt marshes and can be recognized by the concave sides of the stems that create a winged appearance and shorter bracts (1–3 cm long) at the end of the stems.

Origin & Range: Native; scattered locations in Wisconsin; range includes most of U.S.

Habitat: Three-square grows in deep and shallow marshes and along lakes and streams. It is often found in knee-deep water and will grow in water over 1 meter deep on occasion.

Through the Year: Three-square is a perennial that resprouts in the spring from buds formed on the rhizome in the fall. Spikelets form by midsummer and nutlets are mature by late summer.

Value in the Aquatic Community:

A wide variety of ducks including canvasback, gadwall, mallard, pintail, redhead, ringnecked duck and scaup rely on three-square as a food source. Snow geese also feed on it. It is heavily grazed by muskrat and provides cover for waterfowl and other shallow marsh wildlife.

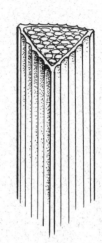

**stem
cross section
4 x life-size**

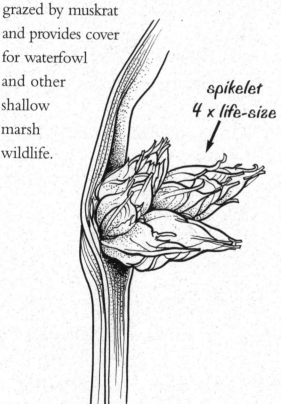

**spikelet
4 x life-size**

A Closer Look:

Snow geese often return to the same three-square beds year after year. A recent study took a look at the impact of repeated heavy grazing on the vigor and nutrient content of the plants. They found that intense grazing did not decrease the nutrient value of the rhizomes, but did create some openings that could be colonized by other plants (Belanger and Bedard 1994).

⅓ life-size

Native

EMERGENT

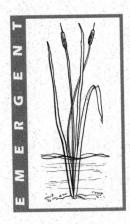

Scirpus fluviatilis (SKER-pus flew-vee-AH-til-es)

River bulrush

Scirpus – (L.) bulrush; *fluviatilis* – (L.) *fluvius* – river

The rain fell in sheets, pelting the river bulrush. The wind shook the sturdy plants. Tall stems, as thick and stout as broom handles, dampened the waves and rain, keeping the soils from entering the lake.

Description: River bulrush stems (7-15 mm thick, up to 2 m tall) are sharply triangular and emerge from a robust, tuber-producing rhizome. The stems have prominent, three-ranked leaves (8-12 mm wide) that are M-shaped in cross-section.

Spikelets (10–25 mm long) are produced on the end of the stem just above several floral leaves. Some of the spikelets are on stalks and others emerge directly from the stem tip. Each spikelet has a spiral of flowers covered by brown scales. Nutlets (4-5 mm long) develop under the scales. Nutlets are three-angled with a prominent beak and six barbed bristles.

Similar Species: River bulrush is part of the sedge family (*Cyperaceae*). It can be separated from grasses and rushes by looking at the spikelets, which have a spiral arrangement of scale-covered flowers and nutlets. It can be separated from other bulrushes in our region by the stout, triangular leafy stem.

Origin & Range: Native; scattered locations, primarily in southern and western Wisconsin; range includes northern and western U.S.

Habitat: River bulrush is found along rivers, lakes, streams and in deep and shallow marshes. It will grow on a range of sediments, from moist shorelines to water over 1 meter deep.

Through the Year: The growth cycle of river bulrush is similar to other regional bulrushes. It overwinters by hardy rhizomes and new shoots emerge in spring. Spikelets are produced by midsummer and mature by late summer. There is significant variability in the number of stalks that produce spikelets from one year to the next.

Value in the Aquatic Community: River bulrush is an excellent shoreline stabilizer. The rhizome network is dense and strong. The rhizomes of river bulrush produce high-nutrient tubers that are an important food source for geese, particularly during migration. Nutlets are eaten by a variety of waterfowl including black duck, canvasback, gadwall, mallard, pintail and redhead. Shoots and rhizomes are consumed by geese and muskrats.

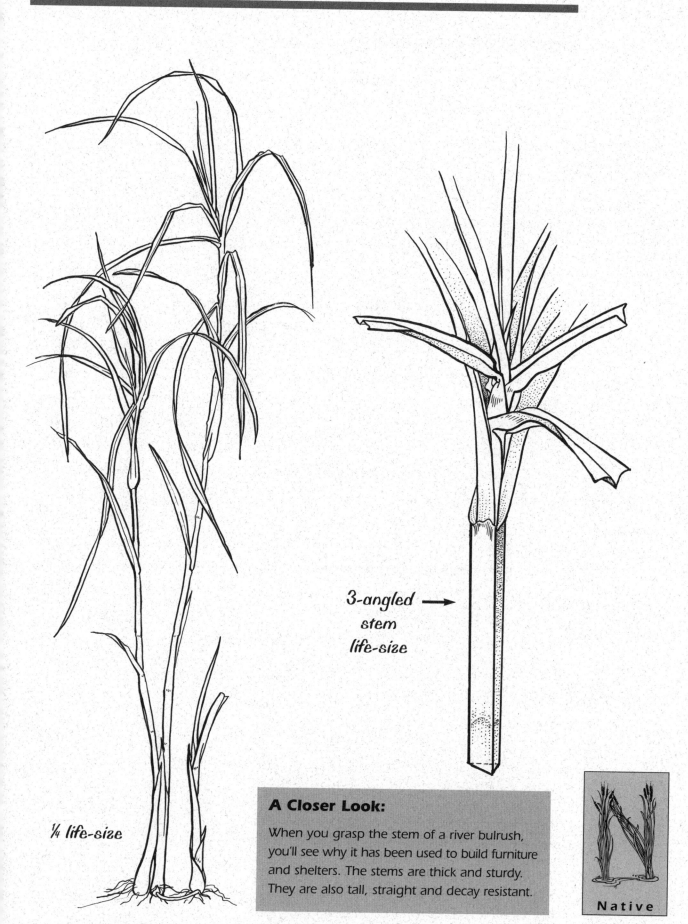

¼ life-size

3-angled
stem
life-size

A Closer Look:

When you grasp the stem of a river bulrush,
you'll see why it has been used to build furniture
and shelters. The stems are thick and sturdy.
They are also tall, straight and decay resistant.

Native

EMERGENT

Scirpus validus (SKER-pus VAL-e-dus)

Softstem bulrush

Scirpus – (L.) bulrush; *validus* – (L.) strong

*The old man and his granddaughter always enjoyed
the adventure of an early morning hike. A twig snapped.
A bittern froze with its bill in the air, camouflaged by the
vertical lines of the softstem bulrush. Its nest was nearby,
built of buoyant softstem shoots.*

Description: Softstem bulrush has tall, flexible stems (1-3 m tall, 1-1.5 cm wide) that emerge from a shallow rhizome. The cylindrical, bluish-green stems are spongy when pressed between your fingers. This is due to the large air chambers that fill the stem. The stems emerge from a slender, buried rhizome and appear to be leafless. However, leaf sheaths are present near the base of each stem. There is also a floral leaf called a bract that looks like a continuation of the tip of the stem. The spikelets emerge in the angle of this bract and the end of the stem.

The oval spikelets, which are about twice as long as wide, are single or clustered on the ends of lax stalks. The scale-covered flowers are spirally arranged. The scales (2.5-3 mm long) are shiny, reddish-brown and often have a green midrib. There may be a few red flecks restricted to the midrib area. Nutlets (1.6-2.1 mm long) develop under the scales.

Similar Species: Softstem bulrush is part of the sedge family. It can be separated from grasses and rushes by looking at the spikelets, which have a spiral arrangement of scale-covered flowers and nutlets. It is most similar in appearance to hardstem bulrush (*Scirpus acutus*). However, hardstem has an olive-green, firm stem filled with smaller chambers than softstem. The spikelets of hardstem are on stiffer stalks and the scales are grayish-brown. The surface of the hardstem scales is also different. They are dull, with shiny red flecks and a fringed margin.

Origin & Range: Native; common throughout Wisconsin; range includes most of U.S.

Habitat: Softstem can be found in wetlands, lakes, ponds and streams. It usually grows in water less than 2 meters deep. Softstem tends to be found on softer, muckier substrates than hardstem and will grow in stagnant water.

lax stalk
1½ x life-size

Through the Year: Softstem bulrush overwinters by hardy rhizomes and dormant winter shoots. Large colonies may develop through rhizome expansion. New shoots sprout in spring and spikelets are formed by midsummer. Nutlets mature by late summer. Some reproduction from seed occurs when exposed mud flats are available for seed germination and seedling development.

Value in the Aquatic Community: Softstem bulrush offers habitat for invertebrates and shelter for young fish. The nutlets are consumed by a wide variety of waterfowl, marsh birds (including bittern, heron, rail) and upland birds. Stems and rhizomes are eaten by geese and muskrats. Bulrushes also provide nesting material and cover for waterfowl, marsh birds and muskrats.

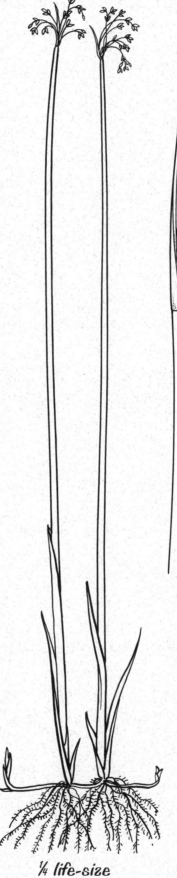

spikelet
4 x life-size

¼ life-size

A Closer Look:

Bulrushes have been used as a food source by many native cultures over the years. The roots were eaten whole or ground into flour. Seeds were used as cereal and the core of young shoots was eaten raw (Whitley et al. 1990).

Native

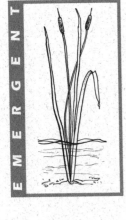

EMERGENT

Sparganium eurycarpum
(spar-GAIN-ee-um yur-ee-CARP-um)

Common bur-reed

Sparganium – (Gk.) *sparganon:* swaddling-band (referring to ribbon-like leaves)
eurycarpum – (Gk.) *eurys:* wide + *carpus:* fruit

A zig-zag stalk of bur-reed juts out of the water. Beneath the pompon-like flowers, a newly emerged damselfly turns into the breeze to air dry its wings. Just below the surface, a gelatinous cluster of salamander eggs coils around the stem.

A Closer Look:

Bur-reed has been used for medicinal purposes over the centuries. The Greek physician Dioscorides recommended a potion of bur-reed root and seed with wine for the treatment of snake bite (Whitley et al. 1990).

Description: There are a number of bur-reeds in Wisconsin, but the common bur-reed (*Sparganium eurycarpum*) is by far the most widespread. Common bur-reed has emergent leaves (6-12 mm wide, up to 1.5 m tall) that are rather spongy and look like a compressed triangle in cross-section. Ribbon-like floating and submersed leaves may also be produced. The leaves and stems sprout from a shallow, spreading rhizome.

The zigzag flower stalk has spherical blooms spaced like sinkers on a fish line. The upper male flowers are the size of small gumballs, while the lower female flowers are the size of jawbreakers. When the female flowers are in bloom, they look like soft, fuzzy balls. When the fruits mature, they have a prickly appearance created by the beaks on the fruits. Each fruit (5-8 mm wide) in the cluster is square-topped with a terminal beak.

Similar Species: Common bur-reed can be distinguished from other bur-reeds by its two stigmas on each female flower and wide, flat-topped fruit. All the other bur-reeds in our region have a single stigma and a narrow, spindle-shaped fruit (see pp. 62-63).

Emergent leaves of common bur-reed are sometimes confused with cattails. A closer look at the leaves shows bur-reed is triangular in cross section and cattail is D-shaped. The floating leaves are sometimes mistaken for wild rice or manna grass. The leaves of bur-reed can be recognized by holding one up to the light to see the very fine checkerboard of veins.

Origin & Range: Native; scattered locations throughout Wisconsin; range includes most of U.S.

Habitat: Common bur-reed is found in marshes and along the margins of lakes, ponds and streams. It grows from moist shoreline soils to water 1 meter deep.

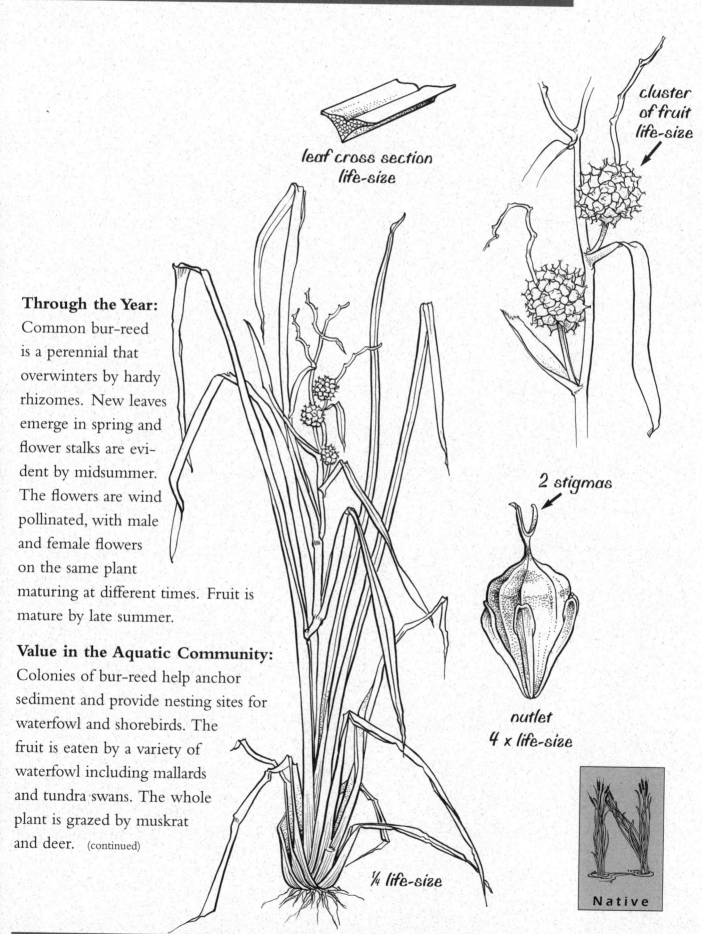

leaf cross section
life-size

cluster
of fruit
life-size

Through the Year:
Common bur-reed
is a perennial that
overwinters by hardy
rhizomes. New leaves
emerge in spring and
flower stalks are evi-
dent by midsummer.
The flowers are wind
pollinated, with male
and female flowers
on the same plant
maturing at different times. Fruit is
mature by late summer.

Value in the Aquatic Community:
Colonies of bur-reed help anchor
sediment and provide nesting sites for
waterfowl and shorebirds. The
fruit is eaten by a variety of
waterfowl including mallards
and tundra swans. The whole
plant is grazed by muskrat
and deer. (continued)

2 stigmas

nutlet
4 x life-size

¼ life-size

Native

Sparganium eurycarpum (continued)

EMERGENT

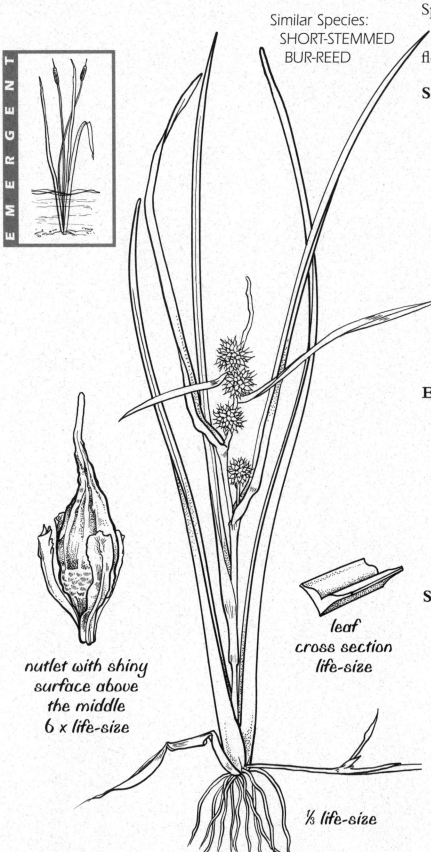

Similar Species:
SHORT-STEMMED
BUR-REED

nutlet with shiny
surface above
the middle
6 x life-size

leaf
cross section
life-size

⅓ life-size

Species that usually have erect leaves, or a combination of erect and floating leaves include:

Short-stemmed bur-reed

(*S. chlorocarpum*) has erect leaves (2-7 mm wide, up to 1 m tall) that are much taller than the flowering stalk. The basal leaves are often in a fan-shaped arrangement. At least some of the fruiting heads (2-2.8 cm diameter) are spaced above the floral bracts. The nutlets have a shiny, greenish-brown surface above the middle. The beak of the nutlet is fairly long (3-5 mm).

Eastern bur-reed

(*S. americanum*) has erect, flat leaves (4-12 mm wide, up to 1 m tall). The fruiting heads (1.5-2.5 cm diameter) are directly in the axils of floral bracts. The nutlets (4.0-5.0 mm long) have a dull, brown surface and a 2.5-4 mm beak.

Shining bur-reed (*S. androcladum*) has erect, keeled leaves (5-12 mm wide, up to 80 cm tall). Their fruiting heads (2-3 cm diameter) are directly in the axils of floral bracts. The nutlets have a shiny, brown surface and a long beak (5-7 mm long).

Species that are usually found with only floating leaves include:

Little bur-reed (*S. minima*) is the smallest bur-reed in our region. The floating leaves are thin and flat (2-7 mm wide). Fruiting heads are only 10-12 mm in diameter and the nutlets each have a tiny beak (0.5-1.2 mm long).

Floating-leaf bur-reed (*S. fluctuans*) has flat, wide floating leaves (5-9 mm wide). The flower stalk is branched with 2-4 fruiting heads (1.5-2 cm diameter). Nutlets are reddish-brown with curved beaks (2-3 mm long).

Narrow-leaf bur-reed (*S. angustifolium*) has floating leaves that are narrower (2-5 mm wide) than those of floating-leaf bur-reed and they are round-ed on the back. The unbranched flower stalk has 1-3 fruiting heads (1.5-1.8 cm diam-eter) with the lowest one usually spaced some dis-tance above the floral leaf (bract). Nutlets have a short, straight beak (1.5-2.5 mm).

Similar Species:
NARROW-LEAF
BUR-REED

spindle-shaped nutlet 6 x life-size

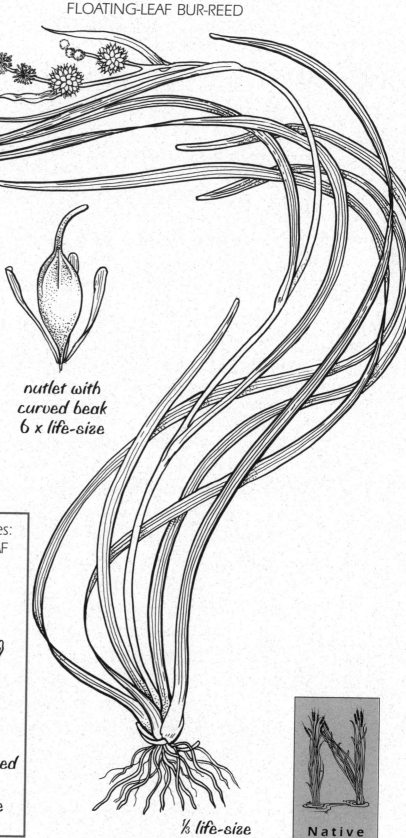

Similar Species:
FLOATING-LEAF BUR-REED

nutlet with curved beak 6 x life-size

⅓ life-size

Native

EMERGENT

Typha angustifolia (TIE-fah an-gus-te-FOL-ee-a)

Narrow-leaved cattail

Typha – (Gk.) *typhe:* cattail; *angustifolia* – (L.) *angusti:* narrow + *folia:* leaves

Narrow-leaved cattails sway slightly in the breeze. The velvety brown spike of the cattail and the song of a red-wing blackbird supply the quintessential elements of a marsh.

Description: Narrow-leaved cattail has dark green, sword-like leaves (5-11 mm wide, 1 m or more tall) that emerge from a robust, spreading rhizome. The leaves are sheathed around one another at the base. At the junction of the leaf sheath and blade, the sheath has membranous ear-shaped lobes called auricles.

The flower looks like a hotdog on a stick. The lower portion is a cylindrical spike (10-20 cm long, 1-2 cm thick) of thousands of tightly-packed female flowers. Each flower has a slender stigma, a fine bract with a spatula-like tip and hairs that are dark on the tips. Some of these female flowers will produce a nutlet and others are sterile. The top of the female spike is separated from the male spike, often by 2 cm or more of bare stem. The male spike has hundreds of anthers that shed pollen to the wind. After the pollen has been released, the male flowers drop off the flower stalk. If you look at the pollen with strong magnification, you can see the pollen grains are individual.

Similar Species: Narrow-leaved cattail can be confused with broad-leaved

cattail (*Typha latifolia*). However, broad-leaved cattail has the male and female flower spikes immediately adjacent to each other and the leaves tend to be wider (10-23 mm wide) and flatter. Details of the flowers are also different. Broad-leaved cattails generally have thicker female spikes (2-3 cm thick). The female flowers each have a broad stigma and no fine bract, and the male flowers produce pollen grains in groups of four. You may also find the hybrid of *T. angustifolia* and *T. latifolia*, known as *Typha* x *glauca*. The hybrid has a blend of features from both species and often has longer female spikes (up to 40 cm long) and taller leaves than either of the parent plants.

Plants without flowers are sometimes confused with blue flag iris (*Iris versicolor*) or sweetflag (*Acorus calamus*). Iris has blue-green leaves arranged in a flat plane. Sweetflag has leaves with an off-center midrib and a spicy smell when crushed.

Origin & Range: Native; common, particularly in southern Wisconsin but the range is increasing with more disturbance; range includes most of U.S.

← *leaf*
sheath
½ life-size

leaf
cross section
2 x life-size

Habitat: Stands of narrow-leaved cattail can be found in marshes, lakeshores, river backwaters and roadside ditches. It will grow in disturbed sites with brackish water up to 0.5 meter or deeper.

Through the Year: Narrow-leaved cattail grows in spring from sprouts produced on the rhizomes during the fall. If mudflats are exposed, seed may have the opportunity to germinate. The flower spikes are produced by mid-summer. The female portion is initially green with a fine sheath. As the flowers mature, the sheath drops and the spike turns brown. After pollen release, the male flowers drop off the flower stalk. Nutlets develop in the female spike by late summer. Each nutlet has a fluff of fine hairs that allows it to be carried by the wind. Some of the seeds disperse in the fall, while others cling to the flower stalk and are not released until the following spring.

Value in the Aquatic Community: Cattails provide nesting habitat for many marsh birds ranging from small (red-winged blackbird, marsh wren) to large (least bittern, coot). Shoots and rhizomes are consumed by muskrats and geese. Submersed stalks provide spawning habitat for sunfish and shelter for young fish.

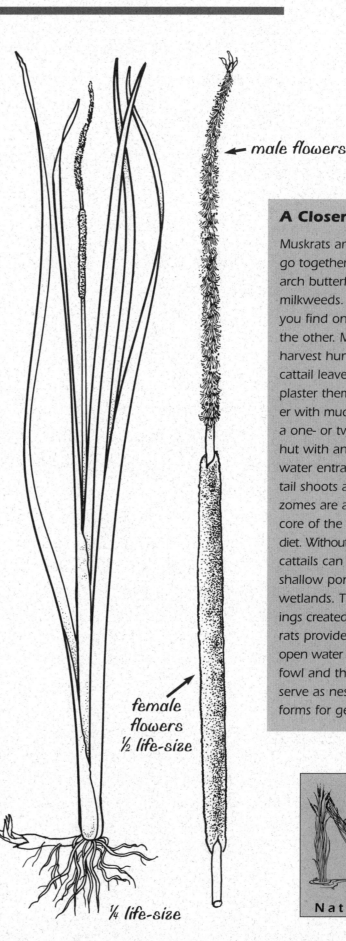

← *male flowers*

female flowers
½ life-size

¼ life-size

A Closer Look:

Muskrats and cattails go together like monarch butterflies and milkweeds. Where you find one, there's the other. Muskrats harvest hundreds of cattail leaves and plaster them together with mud, creating a one- or two-room hut with an underwater entrance. Cattail shoots and rhizomes are also the core of the muskrat diet. Without muskrats, cattails can take over shallow ponds and wetlands. The openings created by muskrats provide essential open water for waterfowl and the huts serve as nesting platforms for geese.

Native

EMERGENT

Typha latifolia (TIE-fah lah-te-FOL-ee-a)

Broad-leaved cattail

Typha – (Gk.) *typhe:* cattail; *latifolia* – (L.) *lati:* broad + *folia:* leaves

The marsh wren knew the route by heart. Weaving through the forest of broad-leaved cattail, she delivered a meal to the small nest filled with ravenous mouths.

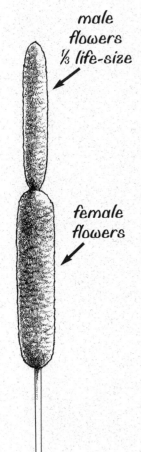

male
flowers
⅓ life-size

female
flowers

Description: Broad-leaved cattail has pale green, sword-like leaves (10-23 mm wide, 1 m or more tall) that emerge from a robust, spreading rhizome. The leaves are sheathed around one another at the base. At the junction of the leaf sheath and blade, the sheath is usually tapered.

The flower looks like a hotdog on a stick. The lower portion is a cylindrical spike (10-15 cm long, 2-3 cm thick) of thousands of tightly-packed female flowers. Each flower has a broad stigma and many white hairs. Some of these female flowers will produce a nutlet and others are sterile. The top of the female spike is immediately adjacent to the male spike, or sometimes separated by a very small space (less than 4 mm). The male spike has hundreds of anthers that shed pollen to the wind. After the pollen has been released, the male flowers drop off the flower stalk. If you look at the pollen with strong magnification, you can see the pollen grains are in groups of four.

Similar Species: Broad-leaved cattail can be confused with narrow-leaved cattail (*Typha angustifolia*). However, narrow-leaved cattail has the male and female flower spikes spaced apart and the leaves tend to be narrower (5-11 mm wide) and rounded on the back. Details of the flowers are also different. Narrow-leaved cattails generally have thinner female spikes (1-2 cm thick), the female flowers each have a slender stigma and a fine bract, and the male flowers produce single pollen grains. You may also find the hybrid of *T. angustifolia* and *T. latifolia,* known as *Typha* x *glauca.* The hybrid has a blend of features from both species and often has longer female spikes (up to 40 cm long) and taller leaves than either of the parent plants.

Plants without flowers are sometimes confused with blue flag iris (*Iris versicolor*) or sweetflag (*Acorus calamus*). Iris has blue-green leaves arranged in a flat plane. Sweetflag has leaves with an off-center midrib and a spicy smell when crushed.

Origin & Range: Native; common in Wisconsin; range includes most of U.S.

Habitat: Stands of broad-leaved cattail can be found in marshes, lakeshores, river backwaters and roadside ditches. It prefers less brackish water than narrow-leaved cattail. It grows from moist soil to water up to a meter deep.

Through the Year: Broad-leaved cattail returns in spring from sprouts produced on the rhizomes during the fall. When there are exposed mudflats, seeds may germinate. The flower spike is produced by midsummer. As the green female flowers mature, the sheath around them drops and the spike turns brown. Pollen from the male flowers is windborne. After pollen release, the male flowers drop off the flower stalk. Nutlets develop in the female spike by late summer. Each nutlet has a fluff of fine hairs that allows it to be carried by the wind. Some of the seeds disperse in the fall, while others cling to the flower stalk and are not released until the following spring.

Value in Aquatic Community:
Cattails provide nesting habitat for many marsh birds. Shoots and rhizomes are consumed by muskrats and geese. Submersed stalks provide spawning habitat and shelter for fish. Invertebrates also live on cattails. The caterpillar of the cattail moth eats seeds of the female flower spike while it produces a network of silky threads that hold the insulating fluff together for over-winter protection (Stokes 1985).

A Closer Look:

Cattails rule many wetlands through their enormous capacity for growth. A single seed can produce a network of rhizomes and a hundred shoots in one growing season. Cattails continue to have many uses in cultures throughout the world. The leaves are dried and woven into mats, chair seats and baskets. Cattail pulp has been used to make paper, and fibers resembling jute can be obtained by treating the leaves and stems with sodium hydroxide (Popenoe et al. 1976). The fluff from the seedheads has served a host of purposes from stuffing life preservers to wall insulation.

The naturalist Euell Gibbons called cattail "the supermarket of the swamp." The core of the rhizomes has been eaten whole or ground into flour. The rhizomes are said to contain as much protein as rice and more carbohydrate than potato (Popenoe et al. 1976). The young shoots have been eaten in salads and the pollen used in baked goods.

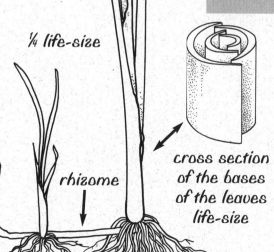

¼ *life-size*

rhizome

cross section of the bases of the leaves life-size

Native

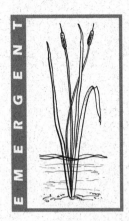

EMERGENT

Zizania spp. (ze-ZAIN-ee-a)

Wild rice

Zizania – (Gk.) *zizanion:* a wild plant of wheat fields

A narrow rice boat maneuvered its way through the bed in a rhythm of poling and knocking. A weathered old man worked the push pole, while the harvester draped ripened rice heads over the bow and knocked the grain off with her ricing stick.

Description: Wild rice is a shallow-rooted annual that sprouts from seed each spring. The first leaves that grow in May and June are narrow, limp and float on the surface of the water. They have a smooth surface and a pointed tip. This is known as the "floating-leaf stage."

By midsummer, flower stalks (2-3 m tall) emerge. These stalks have wide, flat leaves (1-5 cm wide). The flowering portion of the stalk is substantial (10–60 cm). The female spikelets are tightly clustered in a whisk broom-like arrangement on the upper portion of the stalk. (For a definition of spikelet, see discussion under *Glyceria.*) Each female spikelet is one-flowered with a three-ribbed scale called a lemma that extends into a needle-like bristle. Most flowers produce a firm, rod-shaped grain, but some are sterile. Below the cluster of female spikelets, single-flowered male spikelets dangle on spreading stalks.

The two species of wild rice in our region are separated primarily by the appearance of the mature female spikelet.

Southern Wild Rice (*Zizania aquatica*) is usually taller (often 3 m tall with leaves 1-5 cm wide) and more robust than *Z. palustris.* The female spikelet has a dull, papery lemma that is roughened overall. The sterile female spikelets are thread-like (less than 1.5 mm wide). It is the common species in southern Wisconsin rice beds, especially along rivers.

Northern Wild Rice (*Zizania palustris*) varies in size, but is usually shorter than 3 meters with leaves ranging from 4 mm-3 cm wide. The female spikelet has a firm lemma that is smooth between the nerves. The sterile female spikelets are flattened (1.5-2 mm wide). Although *Z. palustris* is smaller in stature, it has larger grains and is therefore commercially more important than *Z. aquatica.* It is the more common species in northern Wisconsin rice beds, especially on lakes. *(continued)*

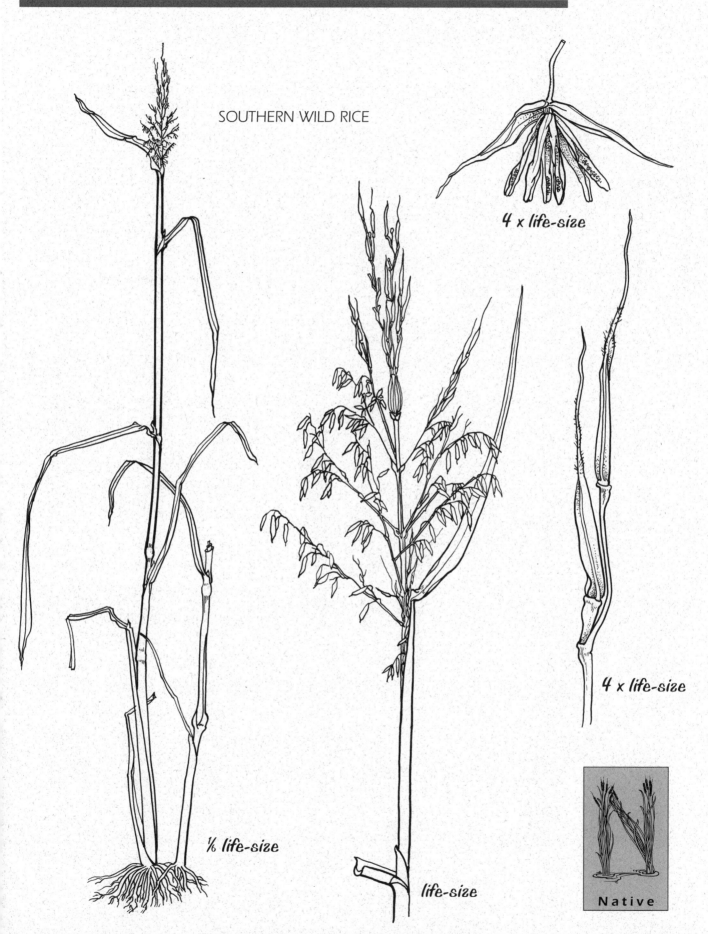

SOUTHERN WILD RICE

4 x life-size

4 x life-size

⅙ life-size

life-size

Native

Zizania spp. (continued)

EMERGENT

NORTHERN
WILD RICE

Similar Species: When wild rice is mature, it is very distinctive and usually not confused with other plants. In the floating-leaf stage, it resembles the floating leaves of manna grass (*Glyceria* sp.) or bur-reed (*Sparganium* sp.). The floating leaves of manna grass have fine hairs on the upper surface, closed leaf sheaths and are rooted with rhizomes. Wild rice has smooth floating leaves, open leaf sheaths and no rhizomes. Bur-reed has floating leaves with rounded tips, no leaf sheaths and a fine checker-board pattern of veins. Wild rice has pointed leaf tips, open leaf sheaths, and longitudinal veins.

Origin and Range: Native; grows throughout the eastern half of U.S. and neighboring portions of Canada, but is most abundant in Minnesota and northern Wisconsin.

Habitat: Wild rice has very specific habitat requirements, including water chemistry. Rice stands usually need a pH of 6.8-8.8, sulfate concentration <10 ppm, and alkalinity from 5-250 ppm. Silt or muck sediment is best, but it will colonize a variety of substrates. Wild rice grows in water depths ranging from 10 cm to 1 meter, but rapid changes in water level can limit success. It is found in streams, rivers and lakes – especially spring-fed lakes or lakes that are part of river systems (Fannucchi et al. 1986).

life-size

⅛ life-size

A Closer Look:

For many years, wild rice was only harvested from natural stands. Over the past 25 years, efforts have been made to domesticate wild rice. Paddy production has been established in California and Minnesota. These paddies are seeded with cultivated varieties bred to hold their seed better for higher yields (Hayes et al. 1989).

The harvest of wild stands is still important to many people, both for economic and cultural reasons. Efforts have been made to protect these rice beds from over-harvest by requiring a license, restricting hours and dates of harvest, and specifying traditional boat and harvest tools. Keeping motor boats out of rice beds and protecting water quality are important. Managing muskrat, beaver and carp can also help improve growth. And part of wild rice success depends on the weather. Gentle summer breezes are best for pollination and stable water levels during June and July help young rice plants take root. Rice harvesters have learned that over a four-year period, there is usually one bumper crop, two fair harvests and one poor yield (Fannucchi et al. 1986).

Through the Year: Seeds germinate in spring and produce floating leaves. Flowering stalks emerge by midsummer. After pollen is shed, the male spikelets start to drop. Grain ripens over a 10-14 day period in late August through mid-September. Grain that is not harvested drops to the water surface and rapidly settles in the soft sediment. Seeds can remain viable for five or more years.

Value in the Aquatic Community: Wild rice is valued by some waterfowl during fall migration. Sora rails and red-wing blackbirds will move to rice beds as the grains mature. Muskrats also use wild rice stems both as a food source and a construction material for lodges.

Native

EMERGENT

Alisma spp. (ah-LIZ-ma)

Water plantains

Alisma – (Gk.) a water plant

*When water plantain is in bloom it has an open airy grace
like the flower baby's breath found in many perennial gardens.
The branched flower stalk towers over the leaves and is
adorned with hundreds of tiny white blossoms.*

Description: Water plantain is in the same family as arrowhead (*Sagittaria* sp.) and has a similar growth form. The leaves emerge from a tuberous base. All the leaves are clustered in a basal rosette (less than 1 m tall) and the flower stalk rises above them. The leaves have an oval-shaped blade and prominent, parallel veins. Each flower is on an individual stalk and has three white petals. The seeds develop in a tightly-packed ring. In deeper water, water plantain will produce long, ribbon-shaped leaves.

There are three species of *Alisma* that may be found in our region.

Southern water plantain (*Alisma subcordatum*) leaves are similar in size and shape to those of northern water plantain, but the flowers are smaller. Each white petal is only

1.8–2.5 mm long. The nutlets have a similar shape to *A. triviale*, but like the flowers they are noticeably smaller (1.5–2.2 mm long).

Northern water plantain (*Alisma triviale*) leaves have a broad, oval blade with a rounded to heart-shaped base. Each flower has three white petals (3.8–4.5 mm long). The nutlets (1.8–3 mm long) have a single groove along their back.

Grass-leaved water plantain (*Alisma gramineum*) has emergent leaves that are shaped like a narrow ellipse. Flower petals (2.3–3.7 mm long) are a pale pink. Each nutlet (1.8–2.6 mm long) has two grooves down its back.

Similar species: The persistent flower stalk of water plantain helps distinguish it from similar-looking emergent arrowheads (*Sagittaria* spp.). When submersed, ribbon-like leaves are produced, a close look is required to separate water plantain from wild celery (*Vallisneria americana*) or submersed arrowhead. The central

SOUTHERN
WATER PLANTAIN

⅙
life-size

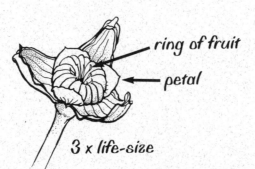

ring of fruit

petal

3 x life-size

stripe of wild celery leaves and the pattern of cross veins in arrowhead help identify them.

Origin & Range: Native; scattered locations, primarily in northern Wisconsin; range includes most of U.S.

Habitat: Water plantain is usually found along the shoreline or in ankle-deep water. Occasionally it will grow as trailing ribbons in deeper zones. It is often found as a single plant or in a small cluster.

Through the Year: Water plantain overwinters by its thick, tuberous rhizome. Reproduction from seed can also occur when conditions are favorable. A new flower stalk develops in early summer and portions of the stalk produce flowers over a number of weeks. Seeds develop by mid- to late summer.

Value in the Aquatic Community: After the flowers go to seed, the sturdy stalk of water plantain becomes a favorite perch for song birds. Both the tubers and nutlets of water plantain are consumed by waterfowl including mallard, pintail, scaup, blue-winged and green-winged teal. The nutlets are also occasionally eaten by pheasants. The foliage offers shade and shelter for young fish. Muskrats and beavers feed on the leaves and roots.

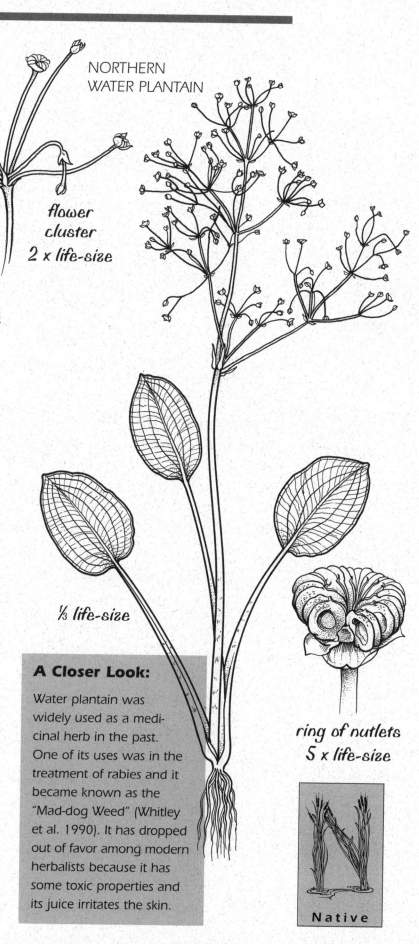

NORTHERN WATER PLANTAIN

flower cluster 2 x life-size

⅓ life-size

ring of nutlets 5 x life-size

Native

A Closer Look:

Water plantain was widely used as a medicinal herb in the past. One of its uses was in the treatment of rabies and it became known as the "Mad-dog Weed" (Whitley et al. 1990). It has dropped out of favor among modern herbalists because it has some toxic properties and its juice irritates the skin.

EMERGENT

Calla palustris (KAH-la pa-LUS-tris)

Wild calla, marsh calla, water arum

Calla – (Gk.) *kollos:* beauty; *palustris* – (L.) of marshes

The elegant, white blooms of wild calla caught the photographer's eye. Kneeling on a clump of sedge, she focused for a close-up of the pale, waxy sheath framed against a deep green heart.

Description: Wild calla has broad, heart-shaped leaves (5-10 cm) that emerge on sturdy stalks (10-20 cm) from a fleshy rhizome. A creamy white floral leaf creates a funnel-shaped cloak around a compact spike of tiny flowers. The fruits are red berries with a few seeds that are crowded on the flower stalk later in the growing season.

Similar species: This shoreline resident is in the same family with jack-in-the-pulpit and sweetflag. The flower structure of wild calla is so unique that it is usually not confused with any other shoreline plant. When not in bloom, wild calla may be mistaken for pickerelweed, which also has heart-shaped leaves but with a rounded rather than pointed tip.

Origin & Range: Native; common in northern Wisconsin; scattered locations in southern portions of state; range includes northern U.S.

Habitat: Wild calla is most often found in shallow water with a peat or muck sediment. It can grow in sprawling chains that send up one leaf at a time, or in compact clusters.

Through the Year: Wild calla overwinters by its hardy rhizome. Reproduction from seed can also occur when conditions are favorable. Flowering starts early in the growing season. Red berries usually develop by mid- to late summer.

Value in the Aquatic Community: The berries of wild calla are consumed by a variety of marsh residents. The leaves and rhizomes are also grazed by muskrats.

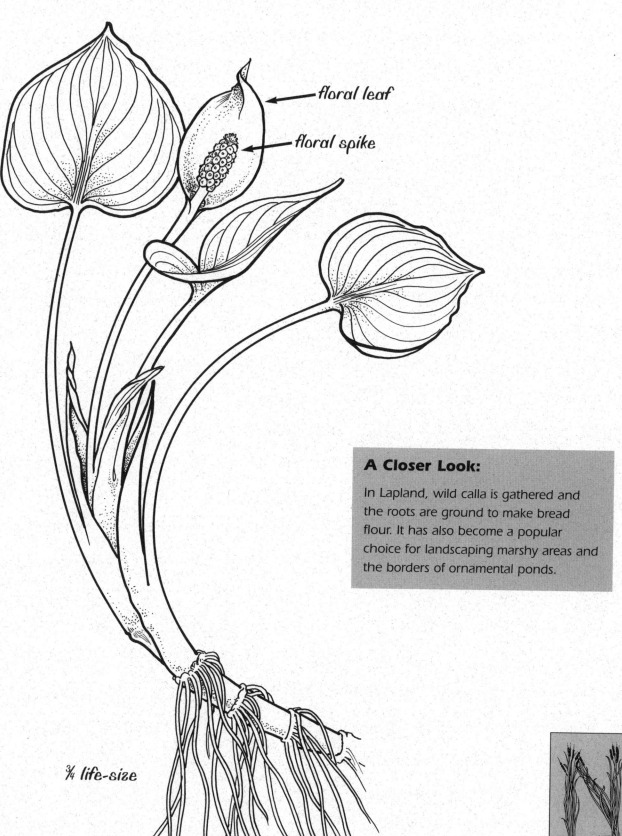

floral leaf

floral spike

A Closer Look:

In Lapland, wild calla is gathered and the roots are ground to make bread flour. It has also become a popular choice for landscaping marshy areas and the borders of ornamental ponds.

¾ life-size

Native

EMERGENT

Cicuta maculata (ce-CUE-ta mac-u-LA-ta)

Water hemlock, spotted cowbane

Cicuta – (L.) poison hemlock; *maculata* – (L.) spotted

The water hemlock flourished in the marshy bay.
The umbrella-like clusters of white flowers suggested its
relation to Queen Anne's lace but did little to divulge
the deadly secret at the root of its bloodline.

Description: Water hemlock has a hollow, erect stem (up to 2 m tall) that emerges from a tuberous rootstalk. Leaves are three-parted, with each part further divided into leaflets (0.5-2 cm wide). The white flowers are in flat-topped, umbrella-like clusters typical of the parsley family (*Umbelliferae*). The fruit is rounded with corky ribs.

Similar species: There is one other species of water hemlock in our region: **bulb-bearing water hemlock** (*Cicuta bulbifera*). It is also poisonous, although not to the same degree as water hemlock.

The leaflets of bulb-bearing water hemlock are narrower (1-5 mm wide) and small

bulblets are produced in the leaf axils.

Origin & Range: Native; scattered locations in Wisconsin; range includes most of U.S.

Habitat: Water hemlock is found growing along moist shorelines of lakes, streams and rivers as well as in marshy areas.

Through the Year: Water hemlock overwinters by its hardy rootstalk. Reproduction from seeds may also occur when conditions are favorable. Flowering occurs by midsummer and fruit develops by late summer.

Value in the Aquatic Community: The fruit of water hemlock is occasionally eaten by marsh birds, but it is usually considered of low importance to wildlife.

narrow
leaflet

Similar Species:
BULB-BEARING
WATER HEMLOCK

⅓ life-size

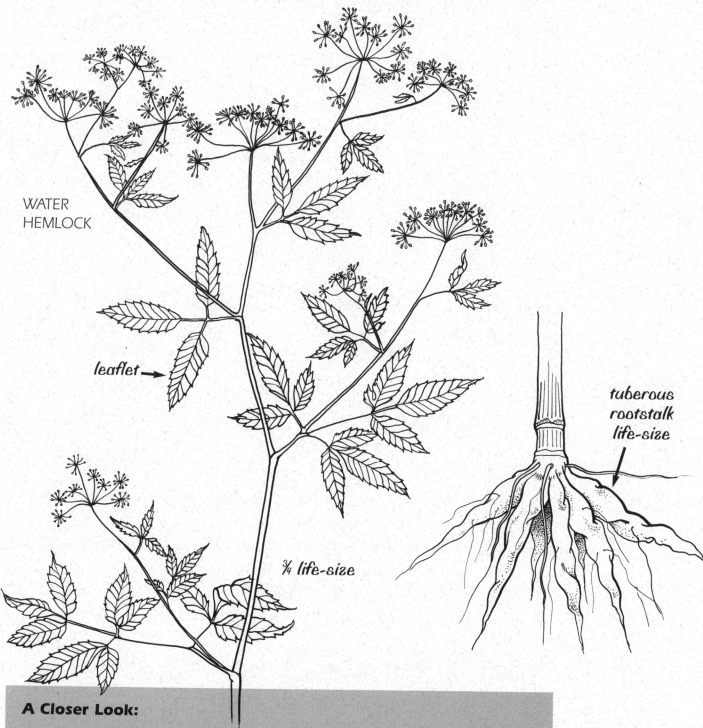

WATER
HEMLOCK

leaflet →

¾ *life-size*

*tuberous
rootstalk
life-size*

A Closer Look:

Anyone with an interest in wild herbs should learn this plant – just so they never eat it! It is considered the most poisonous wild plant in the United States. The toxin, cicutoxin, is concentrated in the tuberous roots but other parts of the plant are also poisonous. Cases of water hemlock poisoning are reported every year for both humans and livestock. The sweet-smelling roots are apparently mistaken for wild parsnips. The toxin is so potent that symptoms appear immediately after consumption. Symptoms include abdominal pain, convulsions, paralysis and respiratory failure, followed by death – sometimes in as quickly as 15 minutes (Voss 1985).

Native

Decodon verticillatus (DECK-o-don ver-TIS-a-LATE-us)

Swamp loosestrife, water-willow

Decodon – (Gk.) *deca:* ten + *odous:* tooth (referring to the toothed margin of the fused sepals); *verticillatus* – (L.) whorled

The spongy stems and arching branches of the swamp loosestrife created the look of a mangrove swamp. The sleek form of a muskrat could be seen swimming among the branches and diving for cover.

Description: Swamp loosestrife has angled, woody stems that emerge from buried rhizomes. The stems may be several meters long with lance-shaped leaves (5-15 cm long, 1-4 cm wide) that are arranged in pairs or whorls of three. Many of the aerial stems arch over and create a tangled network. Thick spongy tissue develops on the underwater stems as well as the enlarged stem tips that root where they dip into the water. Magenta flowers develop in the upper leaf axils. The petals (10-15 mm long) are narrow and slightly crinkled-looking. The fruit (5 mm thick) is a many-seeded capsule.

Similar species: The arching stems with thick, spongy tissue give this plant a unique appearance. The flowers are similar in color and form to those of purple loosestrife (*Lythrum salicaria*) and winged loosestrife (*Lythrum alatum*).

Origin and Range: Native; scattered locations throughout Wisconsin; range includes the eastern U.S.

Habitat: Swamp loosestrife is usually found in shallow water with muck or peat sediment. It can form floating mats in areas of very soft sediment.

Through the Year: New stems emerge from overwintering rhizomes. Flowering usually occurs by midsummer with fruit developing later in the season.

Value in the Aquatic Community: The seeds of swamp loosestrife are grazed by waterfowl including black duck, mallard, blue-winged teal, green-winged teal and wood duck. It can also be a locally important source of food and cover for muskrats.

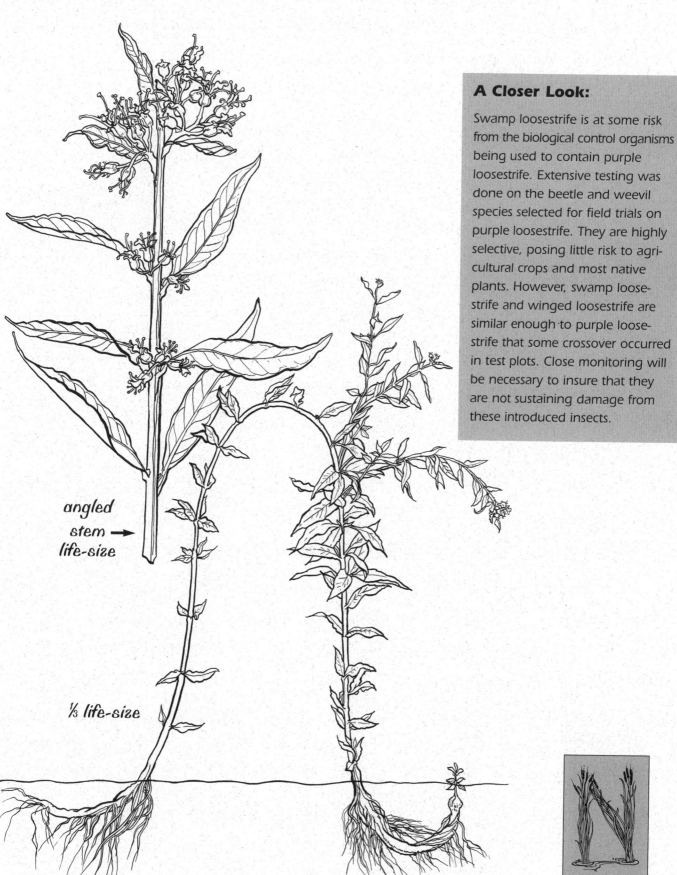

angled
stem →
life-size

⅓ life-size

A Closer Look:

Swamp loosestrife is at some risk from the biological control organisms being used to contain purple loosestrife. Extensive testing was done on the beetle and weevil species selected for field trials on purple loosestrife. They are highly selective, posing little risk to agricultural crops and most native plants. However, swamp loosestrife and winged loosestrife are similar enough to purple loosestrife that some crossover occurred in test plots. Close monitoring will be necessary to insure that they are not sustaining damage from these introduced insects.

Native

EMERGENT

Lythrum salicaria (LITH-rum sal-i-CARE-ee-a)

Purple loosestrife

Lythrum – (Gk.) *lythron*: gore (referring to the color of the flowers);
salicaria – (L.) like a willow

Thickets of purple spikes crowded the wetland that once belonged to native plants. As the cattails disappeared, so did the marsh wrens and least bitterns. The impact of the purple loosestrife invasion was subtle at first and then alarming. What started out as a few stalks of brilliant fall color became a spreading network of fast-growing roots and shoots.

Description: Purple loosestrife has angled stems (50-150 cm tall) that emerge from a woody rootstalk. Leaves (3-10 cm long) are lance-shaped, attach directly to the stem, and often have fine hairs on their surface. The leaves may be opposite, in whorls of three, or sometimes spiraled around the stem. This seems to be related to the number of sides on a stem: four-sided stems have opposite leaves, five-sided stems have leaves in a spiral arrangement, and six-sided stems have leaves in whorls. All three stem types can be found on a single loosestrife plant (Stokes 1985). Clusters of magenta flowers are produced in leaf axils of a terminal spike. Each flower has 5-7 narrow petals (7-12 mm long) that are wrinkled with a tissue paper consistency.

angled stem

life-size

Similar species: Purple loosestrife could be confused with a number of purple-spiked flowers including **gayfeather** (*Liatris pycnostachya*), **blue vervain** (*Verbena hastata*) and **fireweed** (*Epilobium angustifolium*). However, it most closely resembles the other **cultivated purple loosestrife** (*Lythrum virgatum*) and the native **winged loosestrife** (*Lythrum alatum*). *Lythrum virgatum* has smooth leaves and bracts and more flowers per leaf axil than *L. salicaria*. The native winged loosestrife has flowers with shorter petals (4-6.5 mm long) that are solitary in most of the leaf axils.

Origin & Range: Exotic, originated in Europe and temperate regions of Asia. Purple loosestrife was first introduced in the eastern U.S. in the early 1800s. It is currently widespread in Wisconsin. Its range in the U.S. is still primarily in the East and Midwest, but it is becoming a problem in some western locations.

Habitat: Purple loose-strife can be found in a wide variety of sites from moist soil to shallow water. Disturbed sites create an opening for germination of seeds and expansion of new colonies.

Through the Year: Purple loose-strife is a hardy perennial that survives the winter with woody root-stalks. Reproduction from seed is also common throughout the growing season. Seeds have been shown to remain viable in the soil for many years. It only takes about eight weeks from the time of seed germination for a purple loosestrife plant to go into bloom. Flowering usually starts in mid-July and continues through September. Because flowers progressively bloom on the long spike, some seeds are produced earlier in the season than others. Late in the summer, reddish buds are produced at the base of the old flower stalks. These overwinter and produce new flower stalks in the spring. Dead flower stalks remain standing throughout the winter and seeds continue to drop out of dried fruit capsules.

Value in the Aquatic Community: Purple loosestrife has little wildlife value. The seeds are low in nutrition, and the roots are too woody. The flowers are attractive to insects. They produce nectar and are regularly visited by honeybees.

A Closer Look:

Freed from the specialized insects and diseases that keep it in check in its native range of Europe and Asia, purple loose-strife spreads with vigor. Field trials are now being conducted to test European beetles and weevils as potential biological controls. Prevention is the best way to control the spread of purple loose-strife. Pioneer plants should be removed immediately before they have a chance to spread or build up a seed bank. All equipment, clothing and footwear used in areas with purple loosestrife should be thoroughly cleaned before being used at another site. Education is probably the most important strategy. In Wisconsin, it is illegal to sell, distribute, plant or cultivate either *Lythrum salicaria* or *Lythrum virgatum*. There are no exceptions for supposedly "sterile" varieties. A $100 fine is levied for each violation.

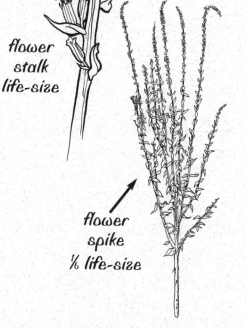

flower stalk life-size

flower spike ⅙ life-size

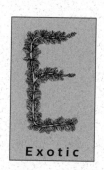

Exotic

EMERGENT

Pontederia cordata (PON-te-DIR-ee-a CORE-da-ta)

Pickerelweed

Pontederia – named for Guilio Pontedera, an Italian botanist (1688–1756);
cordata – (L.) heart-shaped (referring to leaves)

*The huge bed of pickerelweed was humming with business
on the still, humid day. A visiting bee homed in on its goal
using the bright yellow spots on the inside of the blossom.
Each delicate blue flower held a hidden treasure of nectar.*

Description: Pickerelweed has glossy, heart-shaped leaves that emerge from a robust, sprawling rhizome. The leaves have long, air-filled stalks with firm blades. A close look at a leaf blade reveals many fine, parallel veins. The flower spike (5–15 cm long) is crowded with small blue flowers. Each blossom has a tubular portion (5–7 mm long) that contains the nectar and flared lobes (7–10 mm long). The inside of the upper lobe is marked with golden spots. The fruit (5–10 mm long) has a corky, ridged surface and is shaped like an elf's hat. Occasionally plants with white blossoms have been reported. There is also a submersed form that produces narrow, ribbon-like leaves.

Similar species: The blue-flowered spike of pickerelweed is distinctive. When it is not in flower, the leaves might be mistaken for an arrowhead (*Sagittaria* sp.), water plantain (*Alisma* spp.) or wild calla (*Calla palustris*). The heart-shaped base of the blade and the fine, parallel veins are good identifying features.

Origin & Range: Native; common, especially in northern and eastern Wisconsin; range includes eastern U.S.

Habitat: Pickerelweed can be found growing in water from ankle-deep to 2 meters. It will grow in a variety of sediments and often forms spreading colonies in protected bays.

Through the Year: New clusters of leaves emerge in spring from overwintering rhizomes. Flower stalks develop in early summer. Flowers open over time from the bottom of the flower spike to the top. Each flower blooms for only one day and then the upper petals curl inward and a corky, ridged fruit develops. Late in the season, the whole fruiting stalk bends toward the water and seeds can be washed away to new locations. Pickerelweed beds expand both through rhizome growth and reproduction from seeds.

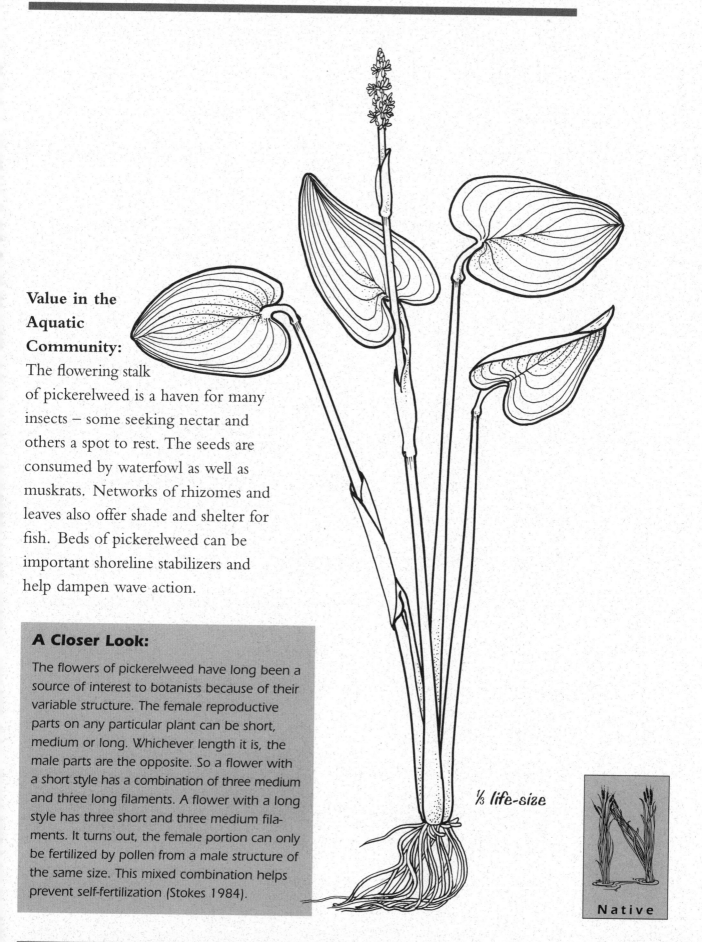

Value in the Aquatic Community:

The flowering stalk of pickerelweed is a haven for many insects – some seeking nectar and others a spot to rest. The seeds are consumed by waterfowl as well as muskrats. Networks of rhizomes and leaves also offer shade and shelter for fish. Beds of pickerelweed can be important shoreline stabilizers and help dampen wave action.

A Closer Look:

The flowers of pickerelweed have long been a source of interest to botanists because of their variable structure. The female reproductive parts on any particular plant can be short, medium or long. Whichever length it is, the male parts are the opposite. So a flower with a short style has a combination of three medium and three long filaments. A flower with a long style has three short and three medium filaments. It turns out, the female portion can only be fertilized by pollen from a male structure of the same size. This mixed combination helps prevent self-fertilization (Stokes 1984).

⅓ *life-size*

Native

E M E R G E N T

Potentilla palustris (PO-ten-TIL-a pa-LUS-tris)

Marsh cinquefoil

Potentilla – (L.) *potens:* potent (referring to medicinal powers);
palustris – (L.) of marshes

*He should have practiced more…the cast toward the boggy
shore snared the trailing stem of the marsh cinquefoil. The
monofilament stretched taut as the maroon flowers bobbed on
the water's surface, startling the insects lured to the plant.*

Description: Although the base of this plant is usually anchored near shore, the stems and leaves often spread out over the water and root from the nodes. Reddish-brown stems grow out from a sprawling rhizome. The leaves have long stalks (20-60 cm) and are divided into 5-7 leaflets. Each leaflet (5-10 cm long, 1-3 cm wide) is shaped like an ellipse with a sharply toothed margin. The flowers (2 cm wide) are a rich reddish-purple. Smooth nutlets develop that are attached to the spongy base of the flower.

Similar species: The leaves of marsh cinquefoil are similar to some cinquefoils that grow on land, but quite different from other aquatic foliage.

Origin & Range: Native; found throughout Wisconsin; range includes eastern U.S. and parts of the west.

Habitat: Marsh cinquefoil is found in the shallow water of swamps, bogs, lakeshores and streambanks. It is usually rooted in muck or peat sediment.

Through the Year: Marsh cinquefoil overwinters as hardy rhizomes and rootstalks. Reproduction from seed may occur when conditions are favorable. Leaves develop in early summer and flowers appear by midsummer. Fruit develops later in the growing season.

Value in the Aquatic Community: The flowers of marsh cinquefoil attract insects. The seeds may be grazed by shore birds or waterfowl. Trailing leaves and stems can provide shade and shelter for fish.

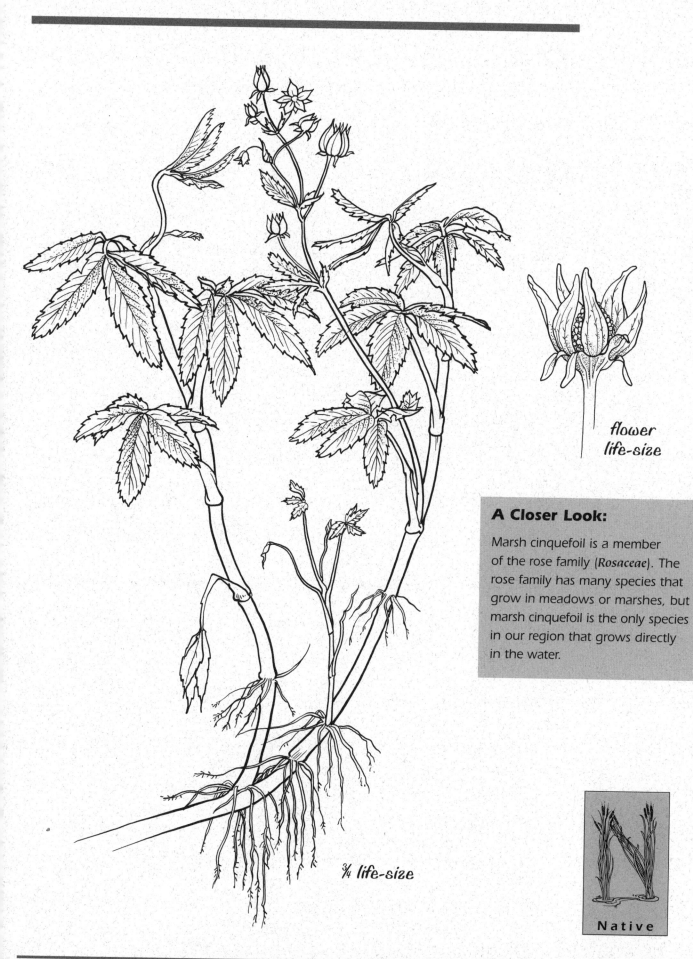

flower
life-size

A Closer Look:

Marsh cinquefoil is a member of the rose family (*Rosaceae*). The rose family has many species that grow in meadows or marshes, but marsh cinquefoil is the only species in our region that grows directly in the water.

¾ life-size

Native

EMERGENT

Rorippa nasturtium–aquaticum (also known as *Nasturtium officinale*)
(ro-RIP-a nas-TUR-shum a-KWA-ti-cum)

Water cress

Rorippa – (L.) marsh cress; *nasturtium* – (L.) *nasus:* nose + *tortus:* to twist
(referring to spicy taste); *aquaticum* – (L.) of the water

*Cool, clear groundwater seeped into the lake along the
gravel shore. It was the perfect home for the bed of water
cress. Beneath the surface, trout foraged on the snails and
scuds that grazed on the cress foliage.*

Description: The spreading stems of water cress are often rooted to the bottom at several points. A single stem may be several meters long and form a tangled mass with other stems. The leaves are subdivided into 3-9 rounded segments, with a terminal lobe that is often larger than the rest. Clusters of white flowers develop on the ends of shoots. Each flower (5 mm wide) has four petals, typical of the mustard family (*Cruciferae*). The fruit is a slender pod (1-2.5 cm long) containing coarsely textured seeds.

Similar species: Water cress resembles bitter cress (*Cardamine* sp.). However, bitter cress doesn't usually form large beds and the stems lack the scattered roots found on water cress. In addition, the seeds of bitter cress are smooth while those of water cress are heavily textured.

Origin & Range: Naturalized; it is thought that water cress was originally introduced from Europe, but it has become so thoroughly naturalized that some botanists consider it a native. Common in Wisconsin. Range includes cold water habitats throughout the U.S.

Habitat: You're unlikely to find this peppery-tasting plant in water warmer than 65°F. Water cress can be found in shallow, sunny spots with cold, clear, running water. It is often found near spring seeps. It will grow in both soft sediment and gravel.

Through the Year: Water cress can remain evergreen when conditions are right. It reproduces both from seed and stem fragments. New foliage develops in spring and flowers form by midsummer. Seeds mature as the growing season progresses.

Value in the Aquatic Community: The leaves of water cress are eaten by waterfowl, muskrats and deer. It is considered an excellent food producer for trout.

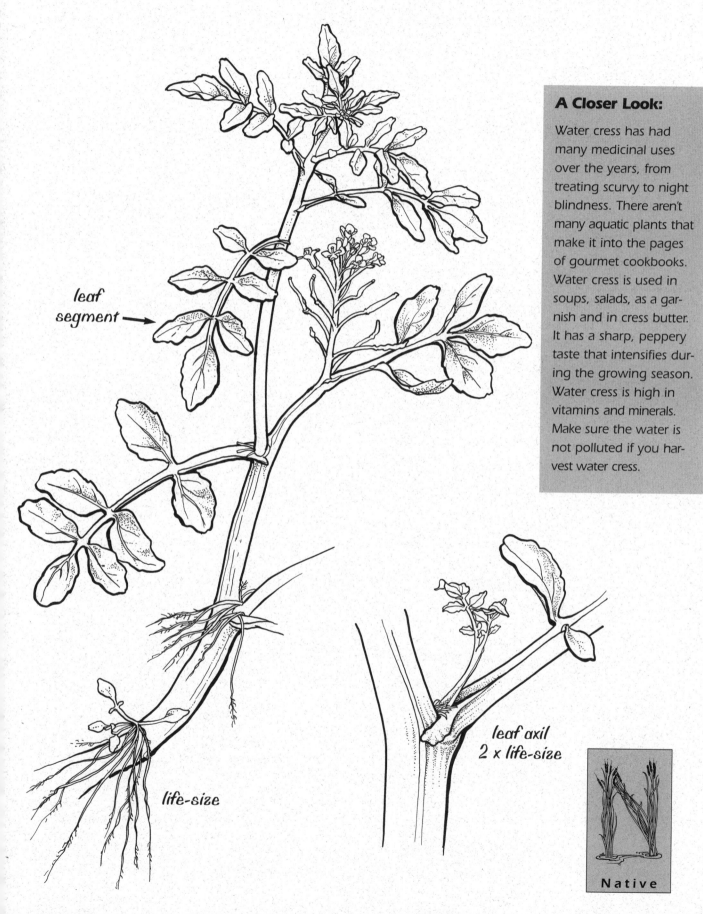

leaf
segment

A Closer Look:

Water cress has had many medicinal uses over the years, from treating scurvy to night blindness. There aren't many aquatic plants that make it into the pages of gourmet cookbooks. Water cress is used in soups, salads, as a garnish and in cress butter. It has a sharp, peppery taste that intensifies during the growing season. Water cress is high in vitamins and minerals. Make sure the water is not polluted if you harvest water cress.

life-size

leaf axil
2 x life-size

Native

Sagittaria graminea (saj-e-TARE-ee-a gra-MIN-ee-a)

Grass-leaved arrowhead, slender arrowhead

Sagittaria – (L.) *sagitta:* an arrow; *graminea:* grasslike

The thunderheads were an omen of things to come. The tiny boat bobbed in the growing swells as the couple scanned the shore. They were looking for the great bed of grass-leaved arrowheads that marked the entrance to the harbor.

Description: Grass-leaved arrowhead has slender leaves (up to 0.5 m tall) that emerge from a short, dense rhizome. The form of emergent leaves varies from almost bladeless to the shape of a wooden spoon. Rosettes of submersed, slender leaves are also often formed. The flowering stem has whorls of slender-stalked male flowers on the upper end and thicker-stalked female flowers below. At the base of each whorl of flowers there are three bracts (3-15 mm long) that are fused for about half their length. The flowers have three rounded, white petals (1-2 cm long). Male flowers have 12 or more stamen with inflated filaments covered with scale-like hairs. Female flowers produce a globe-shaped head packed with dozens of nutlets. Each nutlet (1.2-3 mm long) has a winged margin and sides along with a short, angled beak (< 0.7 mm long).

fused bracts 3 x life-size

Similar species: The leaves of grass-leaved arrowhead are very similar in appearance to those of **stiff arrowhead** (*Sagittaria rigida*). However, the female flowers of stiff arrowhead are either not stalked or have very short stalks, and the main flower stem is strongly bent just above the lowest flowers. The nutlet of stiff arrowhead has a longer beak (0.8-1.4 mm long) than grass-leaved arrowhead.

beaked nutlet with winged margin 10 x life-size

Another species of narrow-leaved arrowhead, **Sagittaria cristata**, is now considered simply a variety of *S. graminea* by most taxonomists. Some of the broad-leaved arrowheads can also produce leaves similar to those of *Sagittaria graminea*. Flower or fruit should be used for identification of species.

Origin & Range: Native; scattered locations, primarily in northern Wisconsin; range includes eastern U.S. *(continued)*

GRASS-LEAVED ARROWHEAD

½ life-size

Native

Sagittaria graminea (continued)

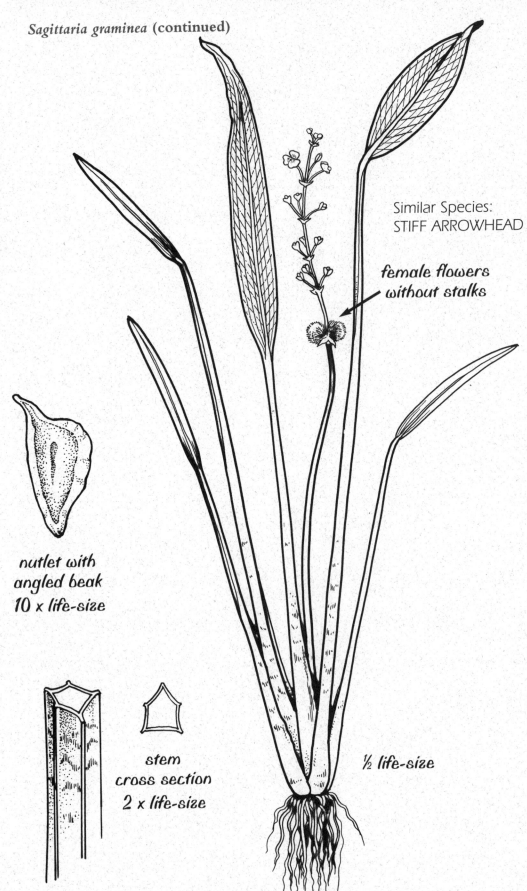

Similar Species:
STIFF ARROWHEAD

*female flowers
without stalks*

*nutlet with
angled beak
10 x life-size*

*stem
cross section
2 x life-size*

½ life-size

Habitat: Grass-leaved arrowhead is found from shallow shorelines to water depths up to 1.8 meters. It will grow in a variety of sediments and often forms extensive beds, especially along rivers and flowages.

Through the Year: Grass-leaved arrowhead survives the winter by its hardy rhizomes. Reproduction from seed may occur when conditions are favorable. Flowering usually occurs by midsummer with nutlets developing by late summer. Foliage begins to yellow and wither early in the fall.

Value in the Aquatic Community: Grass-leaved arrowhead has high wildlife value. Waterfowl graze on the rhizomes and the seeds are consumed by a wide variety of ducks, geese, marsh birds and shore birds. Muskrats, beavers and porcupines eat both leaves and rhizomes. Arrowhead beds offer shade and shelter for young fish.

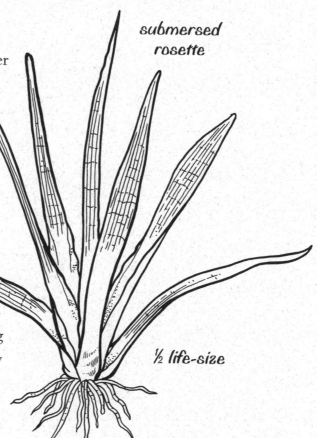

submersed rosette

½ life-size

A Closer Look:

Sagittaria spp. have been an important source of food for people as well as wildlife. Studies of fossilized remains in the Great Basin of the western U.S. revealed that people were eating arrowhead as part of their diet 3000 years ago (Neumann et al. 1989). Arrowhead is still prized by wild food enthusiasts, who prepare the tubers in the same way as potatoes. In China and Japan, arrowhead is cultivated along the margins of rice paddies. Some of these cultivated tubers get as big as oranges (Whitley et al. 1990).

Native

EMERGENT

Sagittaria latifolia (saj-e-TARE-ee-a lah-ti-FOL-ee-a)

Common arrowhead, broad-leaf arrowhead, duck potato, wapato

Sagittaria – (L.) *sagitta:* an arrow; *latifolia* – (L.) *lati:* broad + *folia:* leaves

Fall was harvest time in the arrowhead plots . . . dozens of ducks toiled, bottoms up. They prodded the sediment with their bills. Some tubers were eaten underwater and others were gobbled when they popped to the surface. A muskrat carried a mouthful of tubers to its hidden cache.

Description: Common arrowhead usually produces leaves that are true to its name – shaped like an arrowhead. Leaves emerge in a cluster from tuber-tipped rhizomes. The size and shape of the leaf is highly variable. The blades (5-40 cm long, 0.5-25 cm wide) range from a slender "A" shape to a broad wedge. Some very narrow, knife-like leaves may also be present. The flower stem has whorls of short-stalked male flowers on the upper end and longer-stalked female flowers below. At the base of each whorl of flowers there are three boat-shaped bracts (4–15 mm long). The flowers have three rounded, white petals (1-2 cm long). Male flowers have 20-40 stamen with slender, hairless filaments. Female flowers produce a globe-shaped head packed with dozens of nutlets. Each nutlet (2.5-4 mm long) has a winged margin and a prominent beak (0.6-1.8 mm long) that sticks out at a right angle.

Similar species: Arrowheads have extremely variable leaf shapes. Flowers or fruit are usually necessary to confirm species. There are two other *Sagittaria* spp. in our region that produce broad, arrowhead-shaped leaves.

Arum-leaved arrowhead

(*Sagittaria cuneata*) has leaves similar in shape to common arrowhead, but the lower lobes tend to be much shorter than the upper lobe. The best distinguishing features are the bracts and fruits. The bracts (1-4 cm) are longer and narrower than those of common arrowhead. Each nutlet (1.8-2.6 mm long) has a tiny, upright beak (0.1-0.5 mm long).

Midwestern arrowhead

(*Sagittaria brevirostra*) also has leaves similar to common arrowhead. Key features of Midwestern arrow-head include an angled flower stalk, long bracts (1.4-4 cm long) and nutlets with an upright beak (0.4-1.7 mm long) that is longer than the tiny beak on the fruit of arum-leaved arrowhead. *(continued)*

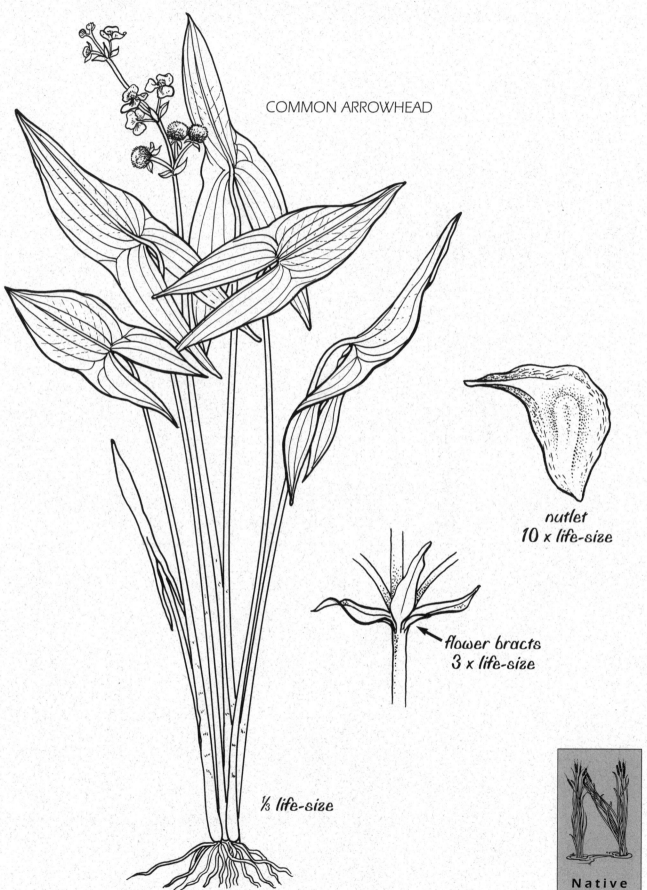

COMMON ARROWHEAD

nutlet
10 x life-size

flower bracts
3 x life-size

⅓ life-size

Native

Sagittaria latifolia **(continued)**

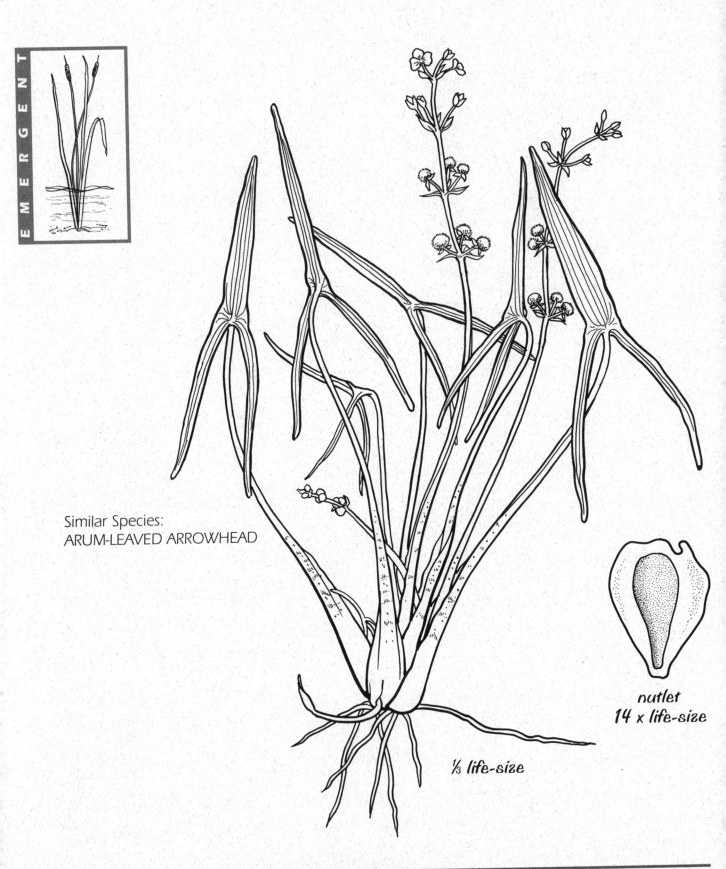

Similar Species:
ARUM-LEAVED ARROWHEAD

nutlet
14 x life-size

⅓ life-size

Origin & Range: Native; common in Wisconsin; range includes most of U.S.

Habitat: Common arrowhead is found in the shallow water of lakes, ponds, streams and marshes. It is usually found in water only ankle-deep, but will sometimes grow in water about 1 meter deep. It can grow in a variety of sediment types.

Through the Year: Common arrowhead survives the winter by its hardy rhizomes and tubers. Reproduction from seed may occur when conditions are favorable. Flowering usually occurs by midsummer with nutlets developing by late summer. Foliage begins to yellow and wither early in the fall.

Value in the Aquatic Community: Common arrowhead is one of the highest value aquatic plants for wildlife. Waterfowl depend on the high-energy tubers during migration and the seeds are also consumed by a wide variety of ducks, geese, marsh birds and shore birds. Muskrats, beavers and porcupines are known to eat both tubers and leaves. Arrowhead beds offer shade and shelter for young fish.

A Closer Look:

Common arrowhead is proving effective in constructed wetlands designed for water quality improvement. It has a wide pH tolerance and can grow on a variety of sediments. It rapidly removes phosphorus from sediments and can store high levels in its leaf tissue. Propagation is relatively easy with tubers, making it a good choice for restorations and wastewater treatment wetlands (Marburger 1993).

Native

Free-floating Plants

"The Earth's vegetation is part of a web of life in which there are intimate and essential relations between plants and the earth, between plants and other plants, between plants and animals. Sometimes we have no choice but to disturb these relationships, but we should do so thoughtfully, with full awareness that what we do may have consequences remote in time and place."

Rachel Carson, 1962

Small duckweed

Great duckweed

Forked duckweed

Common watermeal

Slender riccia

Lemna minor (LEM-na MI-ner)

Small duckweed, water lentil, lesser duckweed

Lemna - (Gk.) a water plant; *minor* - (L.) smaller

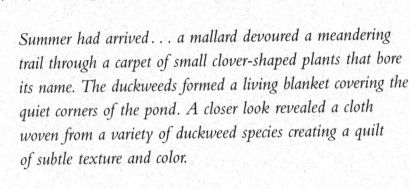

Summer had arrived . . . a mallard devoured a meandering trail through a carpet of small clover-shaped plants that bore its name. The duckweeds formed a living blanket covering the quiet corners of the pond. A closer look revealed a cloth woven from a variety of duckweed species creating a quilt of subtle texture and color.

Description: Small duckweed has round to oval-shaped leaf bodies called fronds (2-6 mm long and 1.5-4 mm wide) that float individually or in groups on the water surface. Each frond has three faint nerves, one root and no stems. The flowers can only be seen with magnification and are seldom found. The plants commonly multiply by budding, remaining attached in small colonies of 2-8 leaf-like plants. Since the plant is free-floating, it must obtain all of its nutrition from the water by absorbing nutrients through its dangling roots and leaf undersurface.

Similar species: Small duckweed can be distinguished from other duckweeds by the size and shape of its fronds and the single dangling root per frond. Forked duckweed (*Lemna trisulca*) is the only other species of *Lemna* commonly found in Wisconsin.

The "rowboat and oars" shape of forked duckweed makes it easy to separate from small duckweed (see *L. trisulca* description). Because it is small and free-floating, small duckweed is easily overlooked and sometimes mistaken for algae.

Origin & Range: Native; common throughout Wisconsin; range includes most of U.S.

Habitat: Small duckweed is often found intermingled with other duckweed species in the quiet waters of bays and ponds. Because it is free-floating, it drifts with the wind or current and is not dependent on depth, sediment type or water clarity. However, there must be adequate nutrients in the water to sustain its growth. Duckweed is often associated with eutrophic waters and can multiply to large populations in these situations.

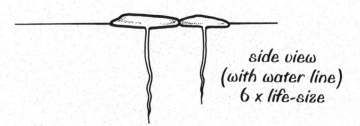

side view
(with water line)
6 x life-size

A Closer Look:

Small duckweed has been used for waste-water treatment. It can remove large amounts of nutrients from the water, has no serious pests, and can be skimmed off periodically. When grown in fertile water, it can reproduce at a tremendous rate – doubling in numbers in three to five days.

Through the Year: Small duckweed survives the winter as small turions or winter buds that rest on the sediment. As the water warms in spring, air spaces develop in the winter buds and they float to the surface. New growth begins and colonies of duckweed expand through the growing season. In late summer and fall, the leaf fronds produce new winter buds that lose buoyancy and sink to the bottom, concluding the annual cycle.

Value in the Aquatic Community: Small duckweed is a nutritious food source that can provide up to 90% of the dietary needs for a variety of ducks and geese. It is also consumed by muskrat, beaver and fish. Rafts of duckweed offer shade and cover for fish and invertebrates. Extensive mats of duckweed can also inhibit mosquito breeding.

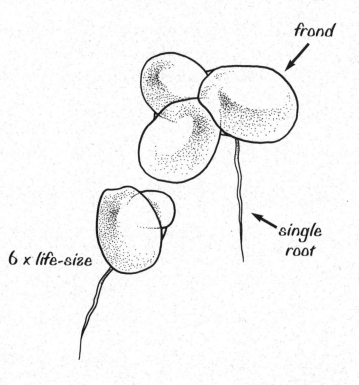

frond

single root

6 x life-size

Native

Lemna trisulca (LEM-na tri-SUL-ka)

Forked duckweed, ivy-leaf duckweed, star duckweed

Lemna – (Gk.) a water plant; *trisulca* – (L.) 3-pointed

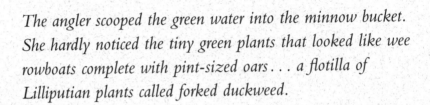

The angler scooped the green water into the minnow bucket. She hardly noticed the tiny green plants that looked like wee rowboats complete with pint-sized oars . . . a flotilla of Lilliputian plants called forked duckweed.

Description: Forked duckweed has a simple, flattened leaf body or frond that is long stalked (4 -16 mm) with three faint nerves and a single root. Lateral fronds (4-10 mm) often remain attached to the parent frond, creating a "rowboat and oars" shape. These angular duckweeds are often hooked together in a tangled mass. Flowers are seldom produced and can only be seen with magnification.

Similar species: Forked duckweed is different from other duckweed species by the stalk-like, "rowboat and oars" shape of the fronds and olive green color.

Origin & Range: Native; found throughout Wisconsin; range includes most of U.S.

Habitat: Forked duckweed is often found just beneath the surface of quiet water. Because it is free-floating, it drifts with the wind or current and is not dependent on depth, sediment type or water clarity. However, there must be adequate nutrients in the water to sustain its growth.

Through the Year: The growth cycle of forked duckweed is similar to other temperate climate duckweeds. It overwinters by producing winter buds that rest on the sediment. In spring the buds become buoyant and float to the surface. Plant growth continues through the summer. Winter buds are formed in the fall and sink to the bottom.

Value in the Aquatic Community: Forked duckweed is a good food source for waterfowl. Tangled masses of fronds also provide cover for fish and invertebrates.

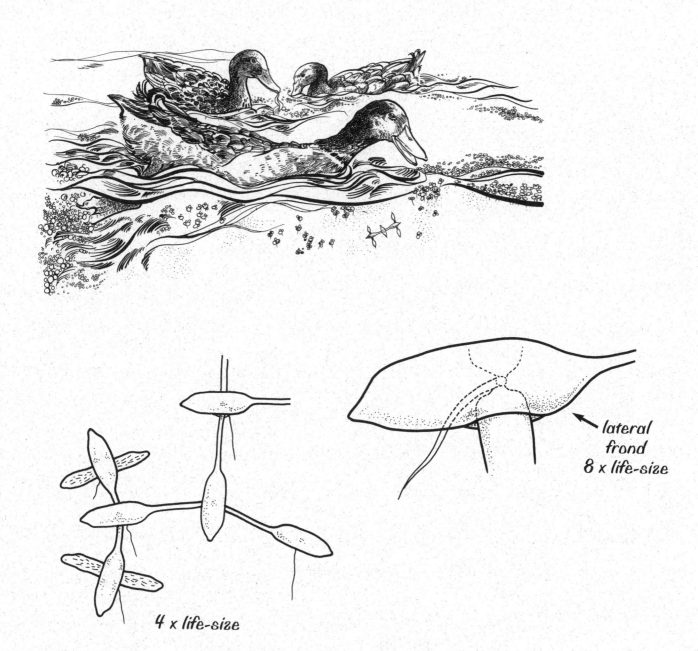

4 x life-size

lateral
frond
8 x life-size

A Closer Look:

Competition studies have been conducted between small duckweed (*Lemna minor*) and forked duckweed. Dense growth of small duckweed can limit the growth of forked duckweed by shading it, because small duckweed grows on the surface while forked duckweed is slightly submersed. However, forked duckweed is more efficient at nutrient uptake with its thin, elongate fronds and greater surface area. In nutrient-limited conditions, forked duckweed becomes a better competitor with small duckweed (Gopal and Goel 1993).

Native

Riccia fluitans (RICH-ee-a FLU-it-ans)

Slender riccia

Riccia – named after P. F. Ricci, an 18th century Italian nobleman;
fluitans – (L.) flowing

The turtle had survived ten years on the lake and grown to enormous proportions. Its armored bulk slipped off the log and dove through the floating mat of riccia that had drifted into the bay.

Description: Mosses and liverworts are usually anchored to a single spot on the surface of a tree or rock. However, this footloose liverwort lives a traveling life, floating just under the water's surface. The forked stems often join together like a floating jigsaw puzzle. Slender riccia is uniformly green with a flat, forking, stem-like body called a thallus. The equally forked thallus resembles a set of miniature, green deer antlers. The ribbon-like divisions are 0.5-1 mm wide and the overall thallus ranges from 10–50 mm long.

Similar Species: Slender riccia could be confused with forked duckweed because it grows in tangled, free-floating masses just beneath the water's surface. Look closely to distinguish the antler-shaped riccia from the rowboat and oar-shaped forked duckweed.

Origin & Range: Native; found throughout Wisconsin; range includes most of U.S.

Habitat: Slender riccia is often found intermingled with duckweed species. It drifts with the wind or current and is not dependent on depth, sediment type or water clarity. However, it needs adequate nutrients in the water to sustain its growth.

Through the Year: Riccia is a non-flowering plant that reproduces from spores produced in capsules. Growth begins in spring from a thallus that survived the winter or with new growth from a germinated spore. Capsules develop on the lower surface of the thallus during the summer. In the fall, growth slows and the plant becomes dormant.

Value in the Aquatic Community: Waterfowl have been known to eat riccia, perhaps because it grows with duckweed. Masses of riccia also provide shade and shelter for fish and invertebrates.

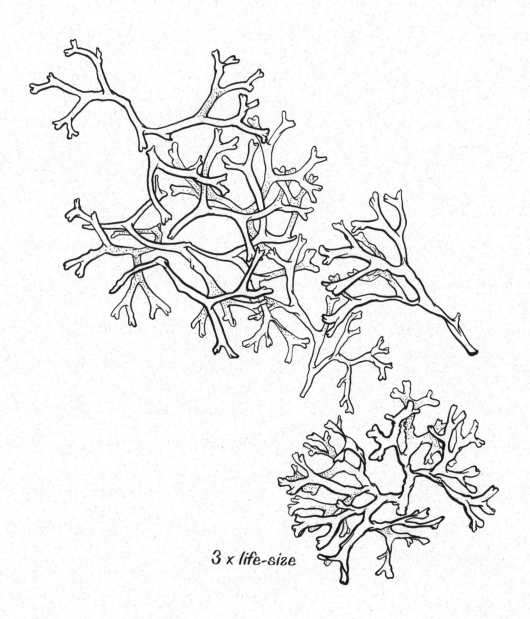

3 x life-size

A Closer Look:

Slender riccia is part of a group of liverworts known as "thalloid" liverworts. They share a common trait of having flattened, branching bodies without leafy structures. These branching patterns can resemble antlers or frost on a window pane.

Native

Spirodela polyrhiza (SPI-ro-DEL-a POL-ee-RIZE-a)

Great duckweed, large duckweed

Spirodela – (Gk.) *speira*: a cord + *delos*: evident;

polyrhiza – (Gk.) *poly*: many + *rhiza*: root

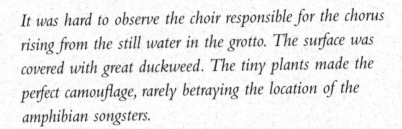

It was hard to observe the choir responsible for the chorus rising from the still water in the grotto. The surface was covered with great duckweed. The tiny plants made the perfect camouflage, rarely betraying the location of the amphibian songsters.

Description: Although great duckweed is the largest of the duckweeds, it is still a very small plant. The simple, flattened "leaf body" or frond has an irregular oval shape (3-10 mm long; 2.5-8 mm wide). Flowers can only be seen with magnification and are seldom found. This plant multiplies mainly by budding. New fronds often remain attached to the parent and create multi-lobed clusters.

Looking up from below the water's surface, great duckweed looks like a tiny sea creature. The green upper surface of each frond has about 5-15 faint nerves radiating from a nodal point. The underside is magenta with a cluster of 5-12 roots dangling down like the tentacles of a miniature jellyfish. The name *polyrhiza* (many roots) describes this feature.

Similar Species: This species can be distinguished from other duckweeds by the size and shape of its fronds, the cluster of multiple roots, and its reddish-purple lower surface.

Origin & Range: Native; found throughout Wisconsin; range includes most of U.S.

Habitat: Great duckweed is often found intermingled with other duckweed species. Because it is free-floating, it drifts with the wind or current and is not dependent on depth, sediment type or water clarity. However, there must be adequate nutrients in the water to sustain its growth.

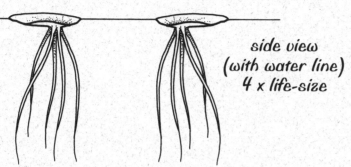

side view
(with water line)
4 x life-size

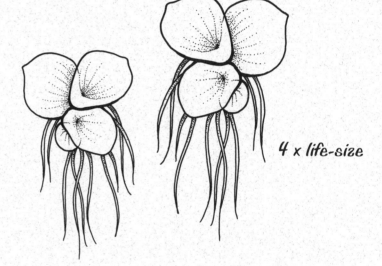

4 x life-size

Through the Year: Great duck-weed is perennial and overwinters by producing kidney-shaped, starch-filled winter buds that look very much like the summer plant, except they are smaller (1-3 mm) and lack roots. These buds sink to the bottom in the fall and float to the surface the next spring, germinating into new plants.

Value in the Aquatic Community: Great duckweed is a good waterfowl food, consumed by many ducks and geese including canvasback, mallard and wood duck. It is also eaten by muskrat and some fish. Rafts of duck-weed offer shade and cover for fish and invertebrates.

A Closer Look:

Both great duckweed and small duckweed are known for their rapid growth and high food value. Studies have shown that a duckweed-covered pond can produce more than twice as much protein per acre as an alfalfa pasture (Whitley et al. 1990). Duckweed has been used effectively as cattle, pig and poultry feed.

Native

Wolffia columbiana (WOLF-ee-a COL-um-bee-AN-a)

Common watermeal

Wolffia – named for J. F. Wolff, a German botanist (1788-1806); *columbiana* – American

FREE-FLOATING

A beaver cuts a black V that opens like a zipper bursting at its seams through a pale green veneer of watermeal. The pint-sized plants, looking like cornmeal, drift into the spaces between other floating plants.

Description: Common watermeal is composed of pale green, asymmetrical globes (0.3-1.4 mm) with no roots, stems, or true leaves. Watermeal has the distinction of being one of the world's smallest flowering plants; however, the flowers can only be seen with magnification and are seldom found. The plants multiply mostly by budding.

Similar Species: This species can be distinguished from other duckweeds by the absence of roots and from other *Wolffia* by its more globular shape, more rounded upper surface, and the absence of dots or markings.

The only other watermeal that is common in our region is **dotted watermeal** (*Wolffia punctata*). It has a football-shaped thallus with a flattened upper surface that is covered with tiny brown dots. Dotted watermeal is usually found growing along with common watermeal. It tends to form a single layer on the water's surface, while common watermeal will sometimes be several layers thick.

Origin & Range: Native; found throughout Wisconsin except the northern lakes and forests ecoregion; range includes most of U.S.

Habitat: Watermeal is often found intermingled with other duckweed species. It drifts with the wind or current and is not dependent on depth, sediment type or water clarity. However, it needs adequate nutrients in the water to sustain its growth.

Through the Year: The growth cycle of common watermeal is similar to other temperate climate duckweeds. It overwinters by producing winter buds that rest on the sediment. In the spring the buds become buoyant and float to the surface. Plant growth continues through summer. Winter buds are formed in the fall and sink to the bottom.

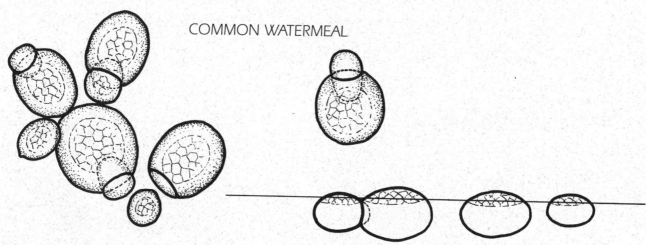

COMMON WATERMEAL

side view
(with water line)
20 x life-size

Value in the Aquatic Community:

Watermeal is good waterfowl food
consumed by a variety of ducks and
geese including mallard and scaup. It
is also eaten by muskrat and some fish.
When large floating rafts form, mosquito
larvae can be prevented from reaching
the surface for oxygen.

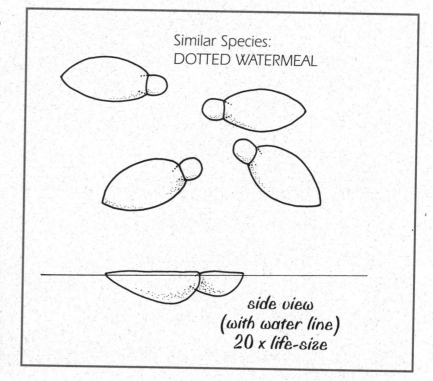

Similar Species:
DOTTED WATERMEAL

side view
(with water line)
20 x life-size

A Closer Look:

A species of watermeal, *Wolffia arrhiza*,
is eaten in Southeast Asia where it is
cooked in soups and stir-fried with other
vegetables. In Thailand, it is called "eggs
of the water" (khai-nam) and raised in
open ponds. Every three to four days it
is skimmed off and taken to market. It is
one of the highest yield crops grown in
Thailand (Popenoe et al. 1976).

Native

Floating-leaf Plants

"A lake is the landscape's most expressive feature. It is earth's eye, looking into which the beholder measures the depth of his own nature."

Henry David Thoreau, 1849

Watershield
American lotus
Yellow pond lily

Spatterdock
White water lily
Water smartweed

Brasenia schreberi (brass-EEN-ee-a SHRE-ber-i)

Watershield, water target

Brasenia – name of unknown origin;
schreberi – named for Johann C. D. Schreber, German botanist (1739-1810)

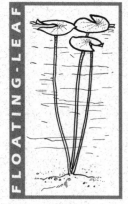

FLOATING-LEAF

He could recognize the watershield with his eyes closed. Most of the plant was encased in a crystal clear gel so slick it slid right through his fingers. Spreading the football shaped leaves, he inspected the thickets of sinuous reddish stems that rose toward the surface.

Description: Stems and leaf stalks of watershield are elastic and allow the floating leaves to ride the waves without uprooting the rhizome that serves as an anchor and source of stored nutrients.

The leaf stalks attach to the middle of the leaves (4-12 cm long, 2-6 cm wide), creating a bull's eye effect that is reflected in the common name "water target." The leaves have a green upper surface and purple underside. Maroon to purple flowers (less than 3 cm wide) are held just above the water surface by stout flower stalks. All submersed portions of the plant are covered with a thick, gelatinous coating.

Similar species: Watershield is sometimes confused with the smaller leaves of water lilies. Lack of a leaf notch and the central location of the petiole help distinguish watershield. Clear gelatinous slime, which coats the stems and underside of the leaves, is also a key characteristic.

Origin & Range: Native; a common species in northern Wisconsin, with scattered locations throughout the state; range includes eastern U.S. and portions of the West.

Habitat: Watershield is most often found in soft-water lakes and ponds, particularly those with sediment that contains partially decomposed organic matter. It grows in water ranging from very shallow depths to about 2 meters.

Through the Year: New shoots develop in the spring from rhizomes, winter buds or seeds. The gelatin-coated leaves unfurl as they reach the water's surface and the upper leaf surface becomes leathery. Flowers are produced in early to midsummer. Watershield flowers have a strategy to ensure cross-pollination by the wind. Each flower is held above the water at two different times: first when the pollen is mature and ready to be released and second, when the female portion of the flower

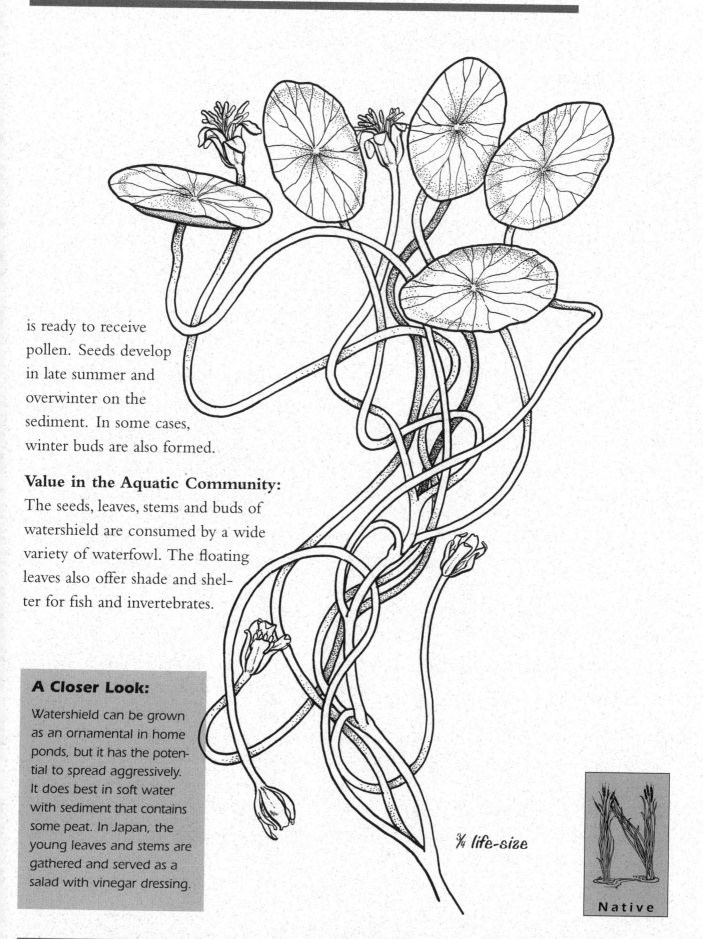

is ready to receive pollen. Seeds develop in late summer and overwinter on the sediment. In some cases, winter buds are also formed.

Value in the Aquatic Community:
The seeds, leaves, stems and buds of watershield are consumed by a wide variety of waterfowl. The floating leaves also offer shade and shelter for fish and invertebrates.

A Closer Look:

Watershield can be grown as an ornamental in home ponds, but it has the potential to spread aggressively. It does best in soft water with sediment that contains some peat. In Japan, the young leaves and stems are gathered and served as a salad with vinegar dressing.

¾ *life-size*

Native

Nelumbo lutea (ne-LUM-bo LOO-tee-a)

American lotus

Nelumbo – from the Sinhalese word nelumbu: Indian lotus; *lutea* – (L.) yellowish

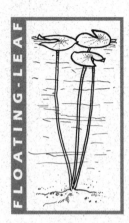

This is a plant of mythic proportions. Its beauty is only surpassed by its legends. Lavish pale yellow blossoms rise above the water, framed by large leaves inverted like umbrellas caught in the wind.

Description: The cylindrical leaf stalks of American lotus emerge from a buried, fleshy rhizome. The stalks support bluish-green leaves that are round, unnotched and very large (30-70 cm in diameter). Some of the leaves may be found floating on the surface, while others are lifted as much as a meter above the water. The leaf stalk is attached to the center of the broad leaves creating a funnel-like appearance. The showy yellow flowers can be the size of dinner plates (15-25 cm wide). A spiral of sepals and petals surround a central seed-bearing structure that looks a bit like a shower head or the top of a watering can.

Similar species: There are only two species of *Nelumbo* in the world: the **sacred lotus** (*Nelumbo nucifera*) of Asia and the American lotus (*Nelumbo lutea*) of the eastern United States. They both have umbrella-like leaves, but the sacred lotus has pink or white flowers, while those of American lotus are yellow. The sacred lotus is cultivated in water gardens and will occasionally show up in our range as an escape. Floating lotus

leaves could be confused with the leaves of white water lily. The water lily has a notched leaf and a larger diameter leaf stalk.

Origin & Range: Native; distribution in Wisconsin includes backwaters of the Mississippi River and on rare occasions inland waters, where it may have been introduced by Native Americans and settlers; range includes eastern U.S.

Habitat: American lotus is usually found in quiet backwaters in water less than 1 meter deep. It can grow in a variety of sediments and will tolerate some turbidity.

Through the Year: New shoots develop in the spring from rhizomes or seeds. Young leaves initially float on the water and then are elevated above it. Flowers form by midsummer. Early in the season, the central portion of the flower, called a receptacle, is spongy and greenish-yellow in color. Developing fruits dot its surface like hard candy pressed into soft dough. As the summer progresses, the petals surrounding the

FLOATING-LEAF

¹⁄₃ *life-size*

receptacle drop off and it becomes brown and dried. The fruits develop thick walls giving them an acorn-like appearance and they rattle in the cavities of the dried receptacle. The flower stalk gradually arches and the receptacle breaks off. It floats upside-down on the surface of the water, gradually scattering fruits and seeds as it breaks down.

Value in the Aquatic Community:

The acorn-like fruit of American lotus is eaten by a variety of waterfowl including canvasback, scaup, mallard and wood duck. The rhizomes are also eaten by beaver and muskrat. Extensive stands of lotus create shade and shelter for fish and wildlife.

A Closer Look:

American lotus was used by several North American tribes as a food source. Fruits were eaten raw or cooked and the tubers were baked. Today, the dried seed pods are commonly used in flower arrangements.

Legends of mystical power and divine wisdom have been associated with both the American lotus and the sacred lotus. The magic of the American lotus is captured in its beauty and the ability of its seeds to sprout after being dormant for as much as 200 years (Whitley et al. 1990).

Native

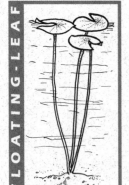

Nuphar advena (NU-far ad-VEEN-a)

Yellow pond lily

Nuphar – (Arabic) *nufar:* water lily; *advena* – (L.) *advenire:* to come to, immigrant

> *The breeze increased throughout the morning. In the once quiet pond, yellow flowers the size of ping-pong balls could be seen bobbing on the surface.*

Description: The leaf and flower stalks of yellow pond lily emerge directly from a robust, spongy rhizome the diameter of a baseball bat. Stalks can grow to be several meters long. Leaves are heart-shaped (20-40 cm long) with rather pointed lobes and have a triangular notch or sinus at their base that looks like it could accommodate a miniature rack of pool balls. Most of the leaves are emergent, growing at an assortment of angles above the water's surface.

Flowers are globular to saucer-shaped (3-5 cm diameter) with five to six yellow sepals (often with a green patch at the base). The sepals curve around many small, strap-like petals, stamen and a yellowish-green disc with the stigmas. This central disc eventually develops into a seed pod.

Similar species: Yellow pond lily most closely resembles the more common spatterdock (*Nuphar variegata*). However, spatterdock has winged leaf stalks, leaves with rounded lobes and sepals with a red rather than green patch.

Origin & Range: Native; distribution in Wisconsin is primarily in the southeastern part of the state; range includes eastern and central U.S.

Habitat: Yellow pond lily is usually found in ponds or slow-moving streams. It can grow in sun or shade, but flowering is more abundant in good light. Yellow pond lily shows a preference for soft sediment and water 2 meters or less deep.

Through the Year: In early summer, clusters of underwater leaves can be seen emerging from the rhizome. As the summer progresses, leaf and flower stalks emerge above the water's surface. Flowering occurs throughout the summer. Flowers open during the day and close at night. The flowers have a fragrance like fermented fruit that attracts insects for pollination. Later in the season, the sepals drop and the central flower structure develops into a fleshy, well-rounded fruit about 4 cm long.

Value in the Aquatic Community: Yellow pond lily provides seeds for waterfowl including mallard, northern pintail, ring-necked duck and scaup. Leaves, stems and flowers are grazed by deer. Muskrat, beaver and porcupine eat the rhizomes. The leaves offer shade and shelter for fish as well as habitat for invertebrates.

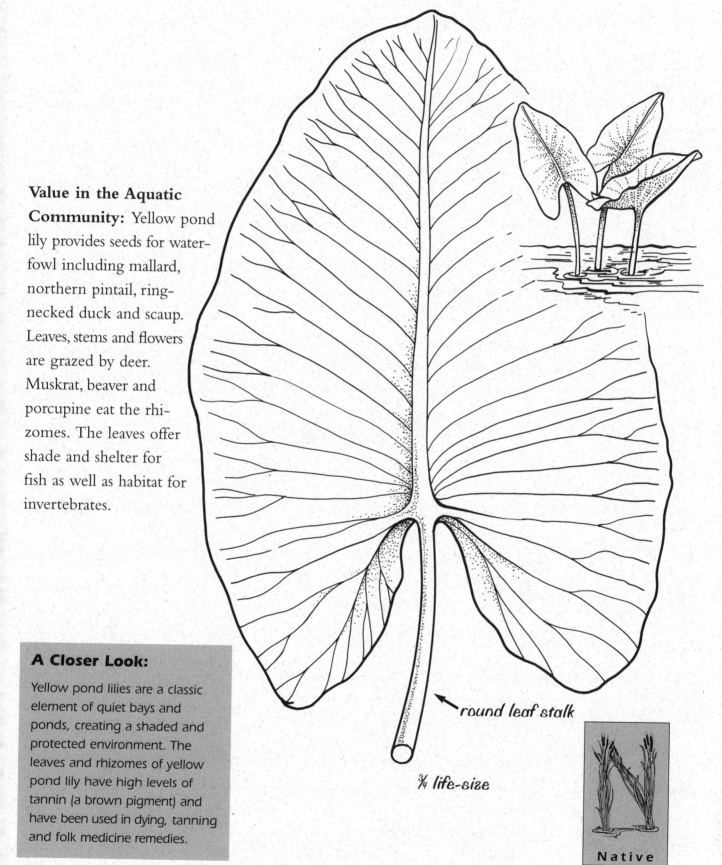

round leaf stalk

¾ life-size

A Closer Look:

Yellow pond lilies are a classic element of quiet bays and ponds, creating a shaded and protected environment. The leaves and rhizomes of yellow pond lily have high levels of tannin (a brown pigment) and have been used in dying, tanning and folk medicine remedies.

Native

Nuphar variegata (NU-far var-ee-a-GAT-a)

Spatterdock, bullhead pond lily

Nuphar – (Arabic) *nufar:* water lily; *variegata* – (L.) varied

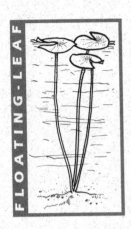

FLOATING-LEAF

Ripples spread across the still water. The deer's attention was focused on the feast at hand. The dinner table was a scene of decapitated flowers stalks and torn leaves . . . all that remained of the bullhead pond lilies.

Description: The leaf and flower stalks of spatterdock emerge directly from a robust, spongy rhizome that is marked with a spiral of scars where old leaf and flower stalks were attached. The sturdy leaf stalks have a flattened upper surface with a narrow wing running down each side. Leaves of spatterdock are heart-shaped (10-25 cm long) with rounded lobes that are parallel or overlapping. The leaf notch is usually less than half the length of the midrib. Most of the leaves float on the water's surface. Flowers are globular to saucer-shaped (2.5-5 cm diameter) with five to six yellow sepals that often have a deep red patch at the base. The sepals curve around numerous small, strap-like petals, stamen and a yellowish-green disc with the stigmas. This central disc eventually develops into a seed pod.

Similar species: Spatterdock most closely resembles yellow pond lily (see the discussion under *N. advena*). Spatterdock may also be confused with *Nuphar microphylla*. The leaves of *N. microphylla*

are small (3.5-10 cm long) with a notch that is two-thirds or more the length of the leaf's midrib. The flowers of *N. microphylla* are also smaller (2 cm or less wide) with a red central disc.

Origin & Range: Native; widely distributed in Wisconsin; range includes eastern and central U.S.

Habitat: Spatterdock is usually found in ponds or slow-moving streams. It can grow in sun or shade, but flowering is more abundant in good light. It shows a preference for soft sediment and water 2 meters or less deep.

Through the Year: In early summer, clusters of underwater leaves can be seen emerging from the rhizome. As the summer progresses, leaves float on the water's surface and flowers rise above them. Flowering occurs throughout the summer. Flowers open during the day and close at night. The flowers have a fragrance like fermented fruit that attracts insects for pollination. Later in the season, the sepals drop and the central

A Closer Look:

Native Americans used both the rhizomes and seeds of spatterdock. The rhizomes were boiled, baked or dried for flour. The seeds were also ground for flour or popped like popcorn.

flower structure develops into a fleshy, well-rounded fruit 2-4.5 cm long.

Value in the Aquatic Community: Spatterdock anchors the shallow water community and provides food for many residents. It provides seeds for waterfowl including mallard, pintail, ringneck and scaup. The leaves, stems and flowers are grazed by deer. Muskrat, beaver and even porcupine have been reported to eat the rhizomes. The leaves offer shade and shelter for fish as well as habitat for invertebrates.

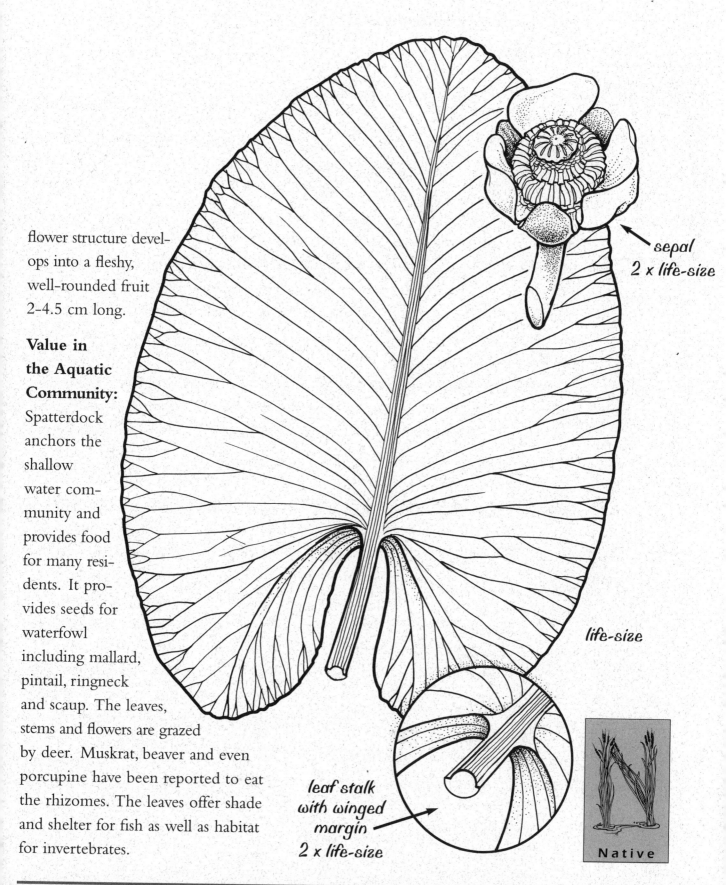

sepal
2 x life-size

life-size

leaf stalk with winged margin 2 x life-size

Native

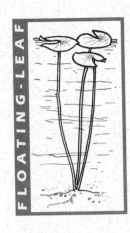

Nymphaea odorata (nim–FEE-a O-dor-AH-ta)

White water lily, fragrant water lily

Nymphaea – (Gk.) nymph; *odorata* – (L.) scented

This is a plant that memories are made of . . . quiet afternoons fishing the edge of water lily meadows; bass and bluegill lingering in their shade during the heat of the day; dragonflies hovering over flowers that release their sweet fragrance on the summer breeze.

Description: The cylindrical leaf stalks of white water lily emerge from a fleshy, buried rhizome. These flexible stalks are round in cross section with four large air passages. The leaves are round (10-30 cm wide) with a narrow sinus and a reddish-purple underside. Most of the leaves float on the water's surface. The flowers (7-20 cm wide) float on the water's surface and are borne on individual flower stalks that arise directly from the rhizome. They have four greenish sepals and numerous white petals in a circular arrangement around the many yellow stamen attached to the central disc.

¾ life-size

Similar species: *Nymphaea odorata* is currently grouped with *Nymphaea tuberosa* by a number of taxonomists. Transplant studies have shown that a number of features that were used to separate these two species were due to growing conditions.

These features included fragrance, time of flower closing and coloration of leaves and flower stalks. Other species that may appear similar to white water lily include **small white water lily** (*Nymphaea tetragona*), yellow pond lily and American lotus. Small white water lily has a northern distribution and is quite rare in Wisconsin. It has smaller, oval leaves (7-12 cm long) with an open sinus. The flowers are also smaller (4-8 cm), lack fragrance and open only in the afternoon. Yellow pond lily and American lotus are easy to distinguish on the basis of leaf shape and flower (see *Nuphar* and *Nelumbo* descriptions). It is important to note that white water lily can produce leaves that are smaller than usual when growing in nutrient-poor conditions or when sprouting from a newly established rhizome.

Origin & Range: Native; widely distributed in Wisconsin; range includes most of U.S.

Habitat: White water lily is usually found in quiet water of lakes or ponds. It grows in a variety of sediment types in water 2 meters or less deep.

Through the Year: In early summer, furled underwater leaves can be seen emerging from the rhizome. As the summer progresses, leaves float on the water's surface with flowers scattered among them. Flowering occurs throughout the summer. Flowers open during the morning and close by midafternoon. When the flower is done blooming, it dips below the water surface where seeds mature inside a fleshy fruit.

Value in the Aquatic Community: White water lily provides seeds for waterfowl. Rhizomes are eaten by deer, muskrat, beaver, moose and porcupine. The leaves offer shade and shelter for fish.

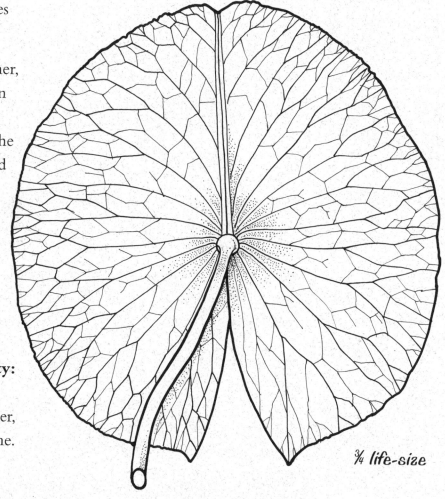

¾ *life-size*

A Closer Look:

White water lily has been widely cultivated for its beauty and wildlife value. It was introduced to English botanical gardens in the late 1700s. Magic powers have been attributed to the white water lily. In medieval times, it was used in love potions. The flower had to be picked during the night of a full moon. The people collecting it had to keep their ears plugged to avoid being bewitched by water nymphs. The dried flower was then worn as a love amulet (Ratsch 1992).

One of the invertebrates often found on water lilies is a beetle in the *Donacia* genus. The female beetle bites holes in the floating leaf and then sticks the lower part of her body through each hole and deposits eggs on the underside of the leaf. When the larvae hatch, they feed on the underwater portions of the lily and suck air from its stem. When it is time for them to spin a cocoon, they attach themselves to the stem and create a watertight case filled with air drawn from the water lily (Stokes 1985).

Native

Polygonum amphibium (po-LIG-o-num am-FIB-ee-um)

Water smartweed, water knotweed

Polygonum – (Gk.) *poly:* many + *gonum:* knee, joint;

amphibium – (Gk.) *amphi:* on both sides + *bios:* life

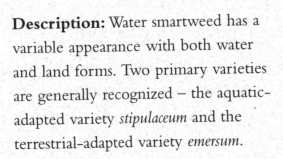

The water smartweed created a sprawling chain of leaves and flowers that reached across the water's surface. The bright pink flowers poked out of the water like candles on a birthday cake.

Description: Water smartweed has a variable appearance with both water and land forms. Two primary varieties are generally recognized – the aquatic-adapted variety *stipulaceum* and the terrestrial-adapted variety *emersum*.

The truly aquatic variety, *stipulaceum* (also known as *P. natans*), shown here, has floating branches with alternate, smooth, elliptical leaves that have a rounded tip. Flowers are arranged in an oval or conical cluster (1.5-4 cm long). When this variety grows on shore it seldom flowers.

The more land-adapted variety *emersum* (also known as *P. coccineum*), has upright stems even when growing in the water and does not produce floating leaves. Leaves of variety *emersum* are hairy and have pointed tips. Flowers are arranged in an extended cylindrical shape that is 4-15 cm long.

Similar species: The floating leaves of water smartweed could be mistaken for the floating leaves of some pondweeds,

such as large-leaf pondweed (*Potamogeton amplifolius*), long-leaf pondweed (*Potamogeton nodosus*) or floating-leaf pondweed (*Potamogeton natans*). However, water smartweed can be easily separated from these pondweeds by its lack of submersed leaves and its swollen nodes on the stem.

Origin & Range: Native; widely distributed in Wisconsin; range includes most of U.S.

Habitat: Water smartweed is usually found in quiet water of lakes, ponds and backwaters. It grows in a variety of sediment types in water less than 2 meters deep.

Through the Year: Water smartweed is a perennial, reproducing by seeds and overwintering rhizomes. New growth emerges from the rhizomes in early summer and flowers develop by midsummer. As the summer progresses, dark, shiny nutlets mature. Late in the growing season, foliage dies back and seeds then drop to the sediment.

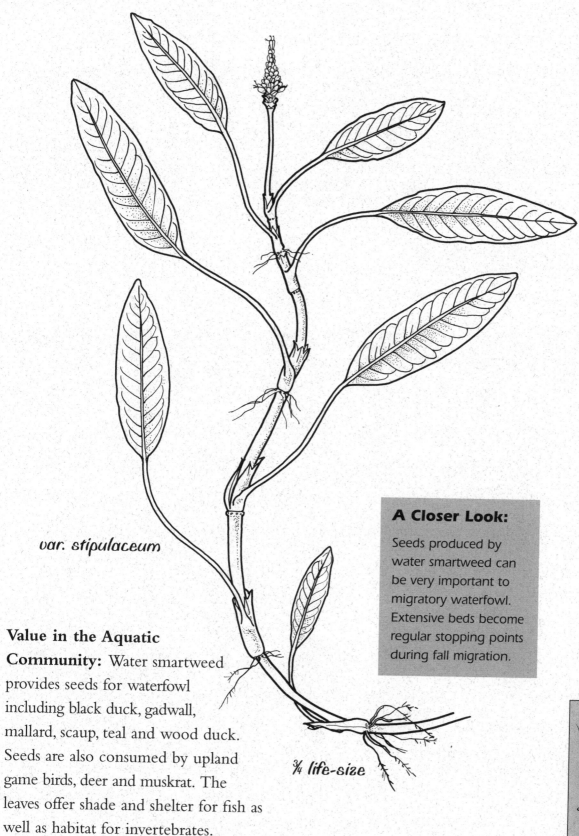

var. stipulaceum

¾ *life-size*

**Value in the Aquatic
Community:** Water smartweed
provides seeds for waterfowl
including black duck, gadwall,
mallard, scaup, teal and wood duck.
Seeds are also consumed by upland
game birds, deer and muskrat. The
leaves offer shade and shelter for fish as
well as habitat for invertebrates.

A Closer Look:

Seeds produced by
water smartweed can
be very important to
migratory waterfowl.
Extensive beds become
regular stopping points
during fall migration.

Native

Submersed Plants

Plants with Entire Leaves

Opposite or Whorled

Water starworts

Muskgrasses

Waterwort

Common waterweed

Golden pert

Slender naiad

Nitellas

Ditch-grass

Horned pondweed

Alternate or Basal

Pipewort

Quillworts

Plantain shoreweed

Water lobelia

Dwarf water milfoil

Pondweeds

Large-leaf pondweed

Algal-leaved pondweed

Curly-leaf pondweed

Water-thread pondweed

Ribbon-leaf pondweed

Leafy pondweed

Variable pondweed

Illinois pondweed

Floating-leaf pondweed

Long-leaf pondweed

Sago pondweed

White-stem pondweed

Small pondweed

Clasping-leaf pondweed

Fern pondweed

Spiral-fruited pondweed

Flat-stem pondweed

Creeping spearwort

Water bulrush

Wild celery

Water stargrass

Plants with Finely-Divided Leaves

Lake cress

Water marigold

Coontail

Farwell's water milfoil

Various-leaved milfoil

Northern water milfoil

Eurasian water milfoil

Stiff water crowfoot

Creeping bladderwort

Large purple bladderwort

Small purple bladderwort

Common bladderwort

"*In the end, we will conserve only what we love… we will love only what we understand… we will understand only what we are taught.*"

Chinese Philosopher,
Lao-Tsu 490 B.C.

Callitriche spp. (cal-LE-tric-key)

Water starworts

Callitriche – (Gk.) *kallitrichos:* beautiful-haired (referring to slender stems)

On the glassy surface of the cool, quiet bay, water striders glided among the delicate green rosettes of water starwort. Each hair-like stem displayed a cluster of rounded leaves that floated on the surface like a bobber on a fish line.

Description: Water starworts are shallowly rooted plants with fine stems, usually less than a meter long. Stems are supported by the water or occasionally sprawled on mud flats. Submersed leaves are opposite, smooth-edged and ribbon-like. Floating leaves (present in all but autumnal water starwort) are more rounded and crowded toward the tip, forming a floating rosette at the surface.

Minute flowers are borne in the axils of the leaves and produce fruit the size of a pinhead. Water starwort species look similar enough that mature fruit is necessary to distinguish them. Three species are found in this region.

Common water starwort (*Callitriche palustris*, formerly known as *C. verna*) is the most common species of water starwort in this area. Slender stems (10-20 cm long) emerge from a shallow rootstalk. Submersed leaves are pale green and linear. The floating rosette has spatula-shaped leaves

(5 mm wide). This plant can be distinguished from the next two species by the appearance of the fruit (1-1.4 mm). Each capsular fruit is slightly longer than wide. The fruit has narrow wings along shallow grooves and surface pitting arranged in vertical rows.

Autumnal water starwort

(*Callitriche hermaphroditica*) has all submersed leaves (5-15 mm long). Leaves are narrow and dark green with a single vein. The 1.5-2 mm fruit is round with a deep groove separating the two halves. There are also winged grooves on the sides of the fruit.

Large water starwort

(*Callitriche heterophylla*) has submersed and floating leaves that are very similar to common water starwort. The fruit of large water starwort is round and just 1 mm wide. The fruit has shallow grooves without wings. The surface has pit-like markings that are not in rows.

A Closer Look:

Water starwort can be successfully cultivated in small pools and aquariums. It needs cool water and plenty of light. Once established, it can be propagated through stem cuttings. The delicate stems and floating rosettes make it a popular choice for ornamental ponds.

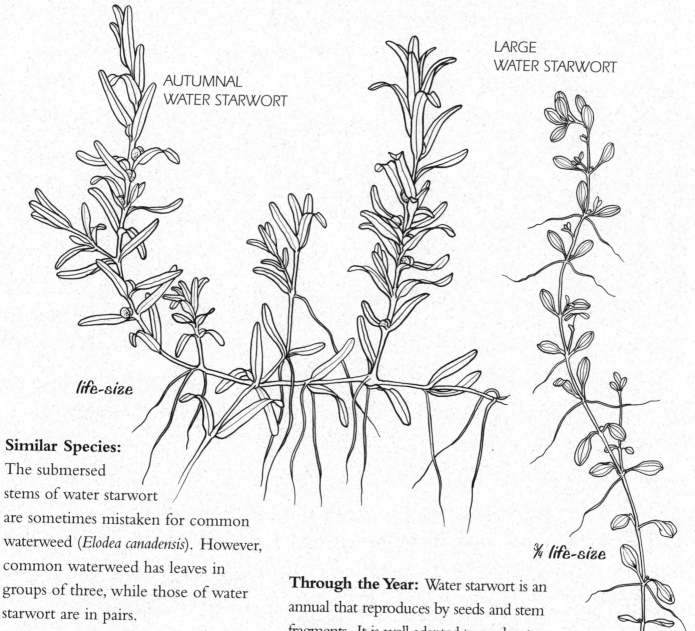

AUTUMNAL
WATER STARWORT

LARGE
WATER STARWORT

life-size

¾ life-size

Similar Species:
The submersed
stems of water starwort
are sometimes mistaken for common
waterweed (*Elodea canadensis*). However,
common waterweed has leaves in
groups of three, while those of water
starwort are in pairs.

Origin & Range: Native;
uncommon in Wisconsin lakes, but
more frequent in streams; autumnal
water starwort and large water starwort
are listed as **Special Concern** species
in Wisconsin. The range of water
starworts includes most of the U.S.

Habitat: Water starwort grows in muddy
or sandy sediment in cool, quiet water.
It is often found in spring-fed waters.

Through the Year: Water starwort is an
annual that reproduces by seeds and stem
fragments. It is well adapted to cool water
and starts growing early in the spring.
Flowers develop in early summer and
seeds mature by mid- to late summer.

Value in Aquatic Community:
The stems and fruit of water starwort
are grazed by a variety of ducks including black duck, bufflehead, canvasback,
gadwall, mallard, redhead and wood
duck. Clusters of stems offer shelter
and foraging opportunities for fish.

Native

Chara spp. (CARR-a)

Muskgrasses, stoneworts, charas

Chara – (L.) a plant

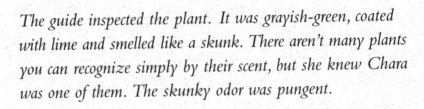

The guide inspected the plant. It was grayish-green, coated with lime and smelled like a skunk. There aren't many plants you can recognize simply by their scent, but she knew Chara was one of them. The skunky odor was pungent.

Description: This unusual type of algae has a growth form that resembles a higher plant, but a close look reveals each joint of the stem is a single cell with no conductive tissue. Muskgrass is simple in structure and has rhizoids rather than true roots. These plants range in size from ankle-high to knee-high. The main branches of muskgrass have ridges. They are often encrusted by calcium carbonate, giving the plant a harsh, crusty feel. The side branches develop in whorls like the spokes of a wheel.

Muskgrass can reproduce vegetatively by spreading rhizoids as well as sexually. The male reproductive structure, called antheridium, and female reproductive structure, called oogonium, are located at the base of branches. Each pear-shaped oogonium is capped with five cells.

Similar species: Muskgrass is similar in appearance to nitella. However, the branches of nitella are smooth and look like they're made of green gelatin, while those of muskgrass are harsh and ridged. With a magnifying glass, you can also see differences in the oogonia: the oogonia of muskgrass have a cap of five cells; those of nitella have ten cells.

Origin & Range: Native; common throughout Wisconsin; range includes most of U.S.

Habitat: Muskgrass is usually found in hard waters. It prefers muddy or sandy substrate and can often be found in deeper water than other plants.

Through the Year: Muskgrass over-winters by rhizoids and fragments. Growth begins when the water warms in spring and continues through the fall.

Value in the Aquatic Community:

Muskgrass is a favorite waterfowl food – more than 300,000 oogonia have been found in the stomach of a single duck. Algae and invertebrates found on musk-grass provide additional grazing. It is also considered valuable fish habitat. Beds of muskgrass offer cover and are excellent producers of food, especially for young trout, largemouth bass and smallmouth bass.

A Closer Look:

The rhizoids of chara slow the movement and suspension of sediments. Therefore, stands of musk-grass can benefit water quality. It is a good bottom stabilizer and is often a pioneer – the first species to invade open areas.

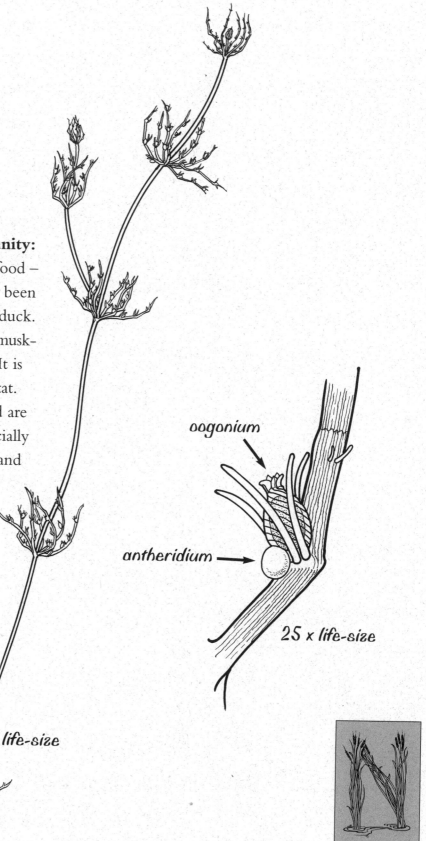

oogonium

antheridium

25 x life-size

life-size

Native

SUBMERSED

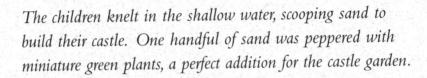

Elatine minima (el–AT–ten–ee MIN–e–ma)

Waterwort

Elatine – (L.) low, creeping plant; *minima* – (L.) smallest

The children knelt in the shallow water, scooping sand to build their castle. One handful of sand was peppered with miniature green plants, a perfect addition for the castle garden.

Description: The whole waterwort plant is not more than a few centimeters tall when it's growing in the water. On exposed mudflats, it forms a low, spreading mat with branches up to 5 cm long. Stems emerge from shallow tufts of roots. The leaves (3-8 mm long) are oblong to oval and attached directly to the stem (no stalk). There is generally a shallow notch at the leaf tip. Flowers have 2-4 sepals and petals and are barely visible in the leaf axils. The capsular fruit is easier to see. It is thin-walled and usually composed of two sections. Inside the capsule, the seeds are all basally attached and stand upright to about the same height. The surface of the seed has an engraved appearance with distinct rows of round to oval shaped pits.

Similar species: There is one other species of waterwort that occurs in Wisconsin: **matted waterwort** (*Elatine triandra*). A close look at the fruits separate these two species. Matted waterwort has three sections per capsule, seeds attached at different levels inside the capsule and six-sided pits on the seed coat. Matted waterwort is listed as a **Special Concern** species in Wisconsin.

Origin & Range: Native; found at scattered locations in the northern lakes and forest ecoregion of Wisconsin; range includes eastern U.S.

Habitat: Waterwort can be found from exposed mudflats out to water several meters deep. It is usually found on sandy sites with low disturbance.

Through the Year: This small annual plant relies on the viability of its seeds

A Closer Look:

Welcome to the world of "Bonsai aquatics" – the really little plants. These diminutive lake inhabitants are often completely missed or mistaken for plant seedlings. Pinhead-sized fruits that form a pod in the leaf axil give away the plant's maturity. Under magnification, you'll notice that the walls of the pods are so thin you can see the seeds inside.

The "signature" created by the pitting pattern on waterwort seeds can only be seen with a microscope, but is a valuable tool for identification. Seeds from a variety of plants have been preserved in sediments for hundreds and even thousands of years. Scientists have been able to identify them by their shape and surface markings.

actual size

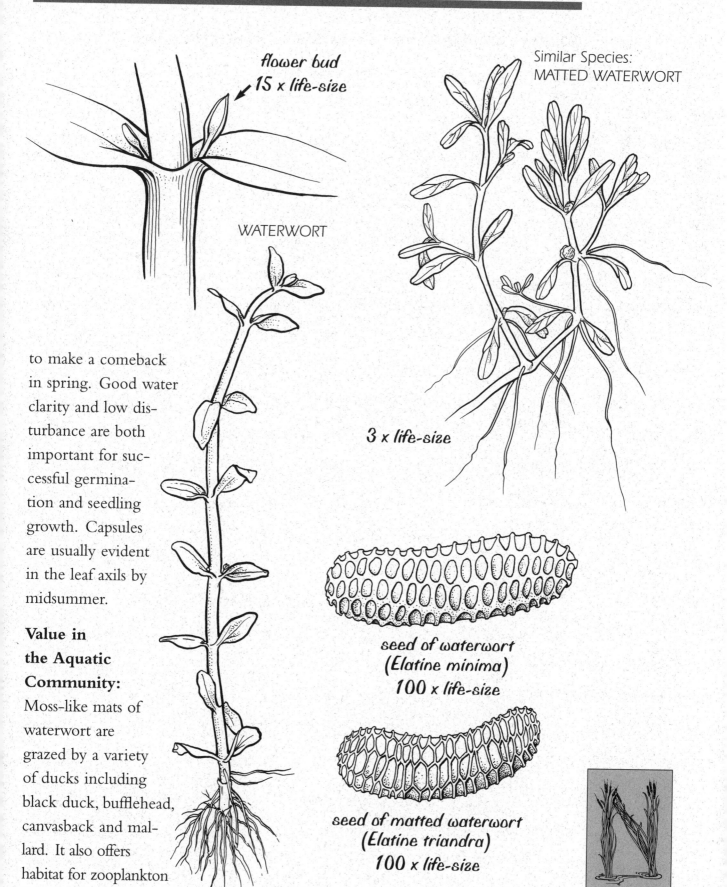

flower bud
15 x life-size

WATERWORT

Similar Species:
MATTED WATERWORT

3 x life-size

to make a comeback in spring. Good water clarity and low disturbance are both important for successful germination and seedling growth. Capsules are usually evident in the leaf axils by midsummer.

Value in the Aquatic Community:

Moss-like mats of waterwort are grazed by a variety of ducks including black duck, bufflehead, canvasback and mallard. It also offers habitat for zooplankton and fish fingerlings.

seed of waterwort
(Elatine minima)
100 x life-size

seed of matted waterwort
(Elatine triandra)
100 x life-size

3 x life-size

Native

SUBMERSED

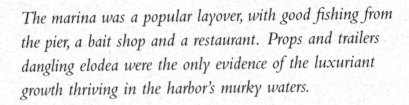

Elodea canadensis (el-oh-DEE-a can-a-DEN-sis)

Common waterweed, elodea

Elodea – (Gk.) *elodes:* marshy; *canadensis:* Canada

The marina was a popular layover, with good fishing from the pier, a bait shop and a restaurant. Props and trailers dangling elodea were the only evidence of the luxuriant growth thriving in the harbor's murky waters.

Description: Common waterweed has slender stems (up to 1 m long) that emerge from a shallow rootstalk. The small, lance-shaped leaves (6-17 mm long, 1-5 mm wide) attach directly to the stem (no leaf stalk). Leaves are in whorls of three, or occasionally only two and tend to be more crowded toward the stem tips. The branching stems often form a tangled mat that can become a nuisance.

Male and female flowers are on separate plants. Female flowers have three small white petals with a waxy surface that improves flotation. They are raised to the surface on a long, slender stalk. Male flowers develop in a vase-like structure called a spathe that is 7-10 mm long. At maturity, the male flowers are also raised to the surface on thread-like stalks. There the anthers split open, releasing pollen to drift away and possibly fertilize female flowers. However, male plants are quite rare. So although you may see dozens of tiny white flowers floating above a bed of common waterweed, they are usually all female flowers that will not produce seed.

Similar species: The only other species of *Elodea* in our region is **slender waterweed** (*Elodea nuttallii*). These two plants look very similar. You need to look at fine details to tell the difference. Slender waterweed is more delicate in structure with finer stems and narrower leaves. The average leaf width for common waterweed is about 2 mm; the leaves of slender waterweed have an average width of 1.3 mm. Leaves of common waterweed tend to be more crowded toward the tip of the stem, whereas the leaves are more evenly spread out on the stems of slender waterweed.

Male flowers are quite a bit more common in slender waterweed and they can help distinguish between these two species. While the male flowers of common waterweed develop in a long, slender spathe (7-10 mm), the spathe

on slender waterweed is compact
and rounded
(only 2-4
mm long).
Male
flowers
of slender
waterweed
break free
from the
spathe at
maturity
and float to
the surface;
those of common
waterweed are
raised on a long stalk.

Origin & Range: Native;
common in Wisconsin; range
includes most of U.S. *(continued)*

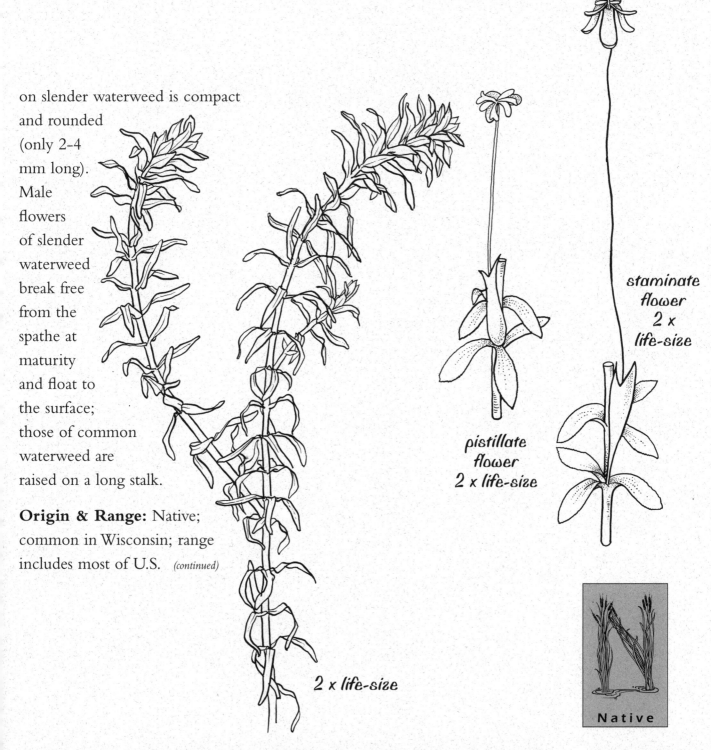

pistillate
flower
2 x life-size

staminate
flower
2 x
life-size

2 x life-size

Native

Elodea canadensis (continued)

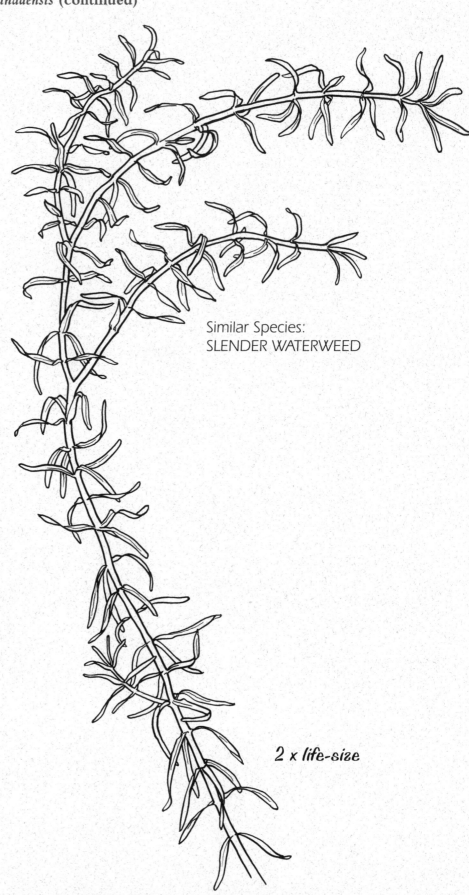

Similar Species:
SLENDER WATERWEED

2 x life-size

Habitat: Common waterweed is found in water depths ranging from ankle deep to several meters deep. It is most abundant on fine sediments enriched with organic matter.

Through the Year: Common waterweed often overwinters as an evergreen plant. Photosynthesis continues at a reduced rate under the ice. In the spring, fresh green shoots develop on the ends of stems. Flowering occurs by early to midsummer. Since seeds are rarely produced, the plant spreads primarily by stem fragments.

Value in the Aquatic Community: The branching stems of common waterweed offer valuable shelter and grazing opportunities for fish, although very dense stands can obstruct fish movement. It also provides food for muskrats and waterfowl. They can eat the plant itself or feed on a wide variety of invertebrates that use the plant as habitat.

A Closer Look:

The success of common waterweed can be attributed to many factors including disease resistance and a tolerance for low light conditions.

In Europe, *Elodea canadensis* is considered an aggressive exotic and is the target of nuisance control programs. Europeans call it "American waterweed." Its ability to spread by stem fragments and tolerate low light have made it a formidable invader.

Elodea also has a "big brother" known as giant elodea (*Egeria densa*). This southern elodea has larger leaves and there are 4-6 leaves in each whorl. It was originally introduced from South America as an aquarium plant, but now grows at nuisance levels in many southern lakes, ponds and ditches.

Native

Gratiola aurea (grah-TEE-oh-la ORE-ee-a)

(formerly known as *Gratiola lutea*)

Golden pert, dwarf hyssop

Gratiola – (L.) *gratia*: grace (referring to healing properties); *aurea* – (L.) golden

The first rays of dawn touched the shallow shore of a northern lake. A blue heron stood motionless, its long legs like pilings amid the golden pert. Beneath the heron, a school of minnows swam among these tiny members of the snapdragon family.

Description: The submersed form of golden pert is only about 1 cm tall. The stems emerge from a cluster of roots. Each stem has opposite, pointed leaves (up to 5 mm long). When growing on shore, the plant is ankle to knee high with glandular, opposite leaves (1 cm long). Shoreline plants produce tubular, yellow flowers in the axils of the leaves.

Similar species: The dwarf, submersed form of *Gratiola aurea* could be confused with waterwort (*Elatine minima* or *Elatine triandra*). However, the leaves of waterwort are more rounded and the flowers or fruiting capsules are usually evident in the leaf axil, whereas submersed golden pert is sterile.

Origin & Range: Native; found at scattered locations in northern lakes; range includes eastern U.S.

Habitat: This tiny member of the snapdragon family is usually found on sandy shores or shallow water of northern lakes. The truly aquatic form is a dwarf, sterile plant that could be mistaken for a seedling. Golden pert can be found on sandy and peat shores of soft water lakes. The submersed form has been found at depths up to 4 meters.

Through the Year: Golden pert is a perennial that overwinters by its root stalk. Shoreline plants can also reproduce from seed. Flowering occurs in midsummer and fruit is produced by late summer.

Value in the Aquatic Community: Moss-like beds of golden pert can provide grazing opportunities for waterfowl. Stands in deeper water also offer foraging for fish.

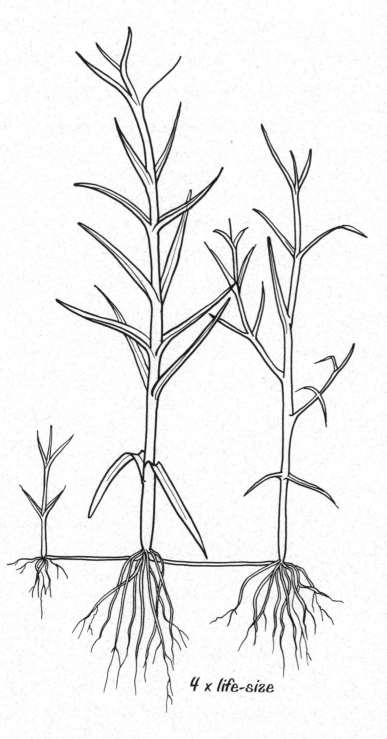

A Closer Look:

The genus *Gratiola* is known for its medicinal values. In the past, extracts of *Gratiola* spp. have been used for a variety of conditions ranging from fever to dropsy. The Latin name *Gratiola* means "grace of god," and was given to this plant for its significance as an herbal medicine (Whitley et al. 1990).

life-size

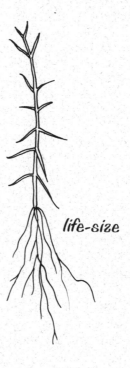

4 x life-size

Native

SUBMERSED

Najas flexilis (NAJ-es FLEX-il-is)

Slender naiad, bushy pondweed

Najas – (Gk.) *naias:* a water-nymph; *flexilis* – (L.) flexible

The flexible stems of slender naiad danced in the waves like their namesake – the water nymphs. The ancient Greeks and Romans believed special nymphs protected the various parts of nature: oreads guarded the hills, dryads took care of forests and naiads gave life and vitality to lakes, rivers and springs.

Description: Slender naiad has fine, branched stems (up to 1 m long) that emerge from a slight rootstalk. The leaves are paired, but there are sometimes bunches of smaller leaves crowded in the leaf axils. Size and spacing of the leaves is extremely variable, depending on growing conditions. Sometimes the plant is compact and bushy, other times trailing and slender. Leaves are narrow with a broad base where they attach to the stem. This base is shaped like sloping shoulders. Each leaf (1–4 cm long, 0.2–1.0 mm wide) tapers to a pointed tip. The leaf margin is finely serrated. Tiny flowers develop in the leaf axils and produce fruit with a paper-thin wall. The seed (2.5–3.7 mm) has a glossy surface with 30–50 rows of small, faint pits.

Similar species: There are three other species of *Najas* that might show up in our region, but *Najas flexilis* is by far the most common.

Southern naiad (*Najas guadalupensis*) has leaves that are wider (up to 2 mm) and less pointed than *N. flexilis.* The seed (1.2–2.5 mm) of southern naiad is dull and deeply engraved with 20–40 rows of angled pits. Southern naiad has a wide distribution but is less common than *Najas flexilis* in northern zones.

Similar Species: NORTHERN NAIAD

Northern naiad (*Najas gracillima*) has thread-like leaves that have a jagged, lobed base. The light brown seed (2.0–3.2 mm) has 20–45 rows of pits that appear stretched. Northern naiad is usually found in soft-water lakes. It is sensitive to pollution and has disappeared in some areas. *(continued)*

jagged lobe at leaf base

10 x life-size

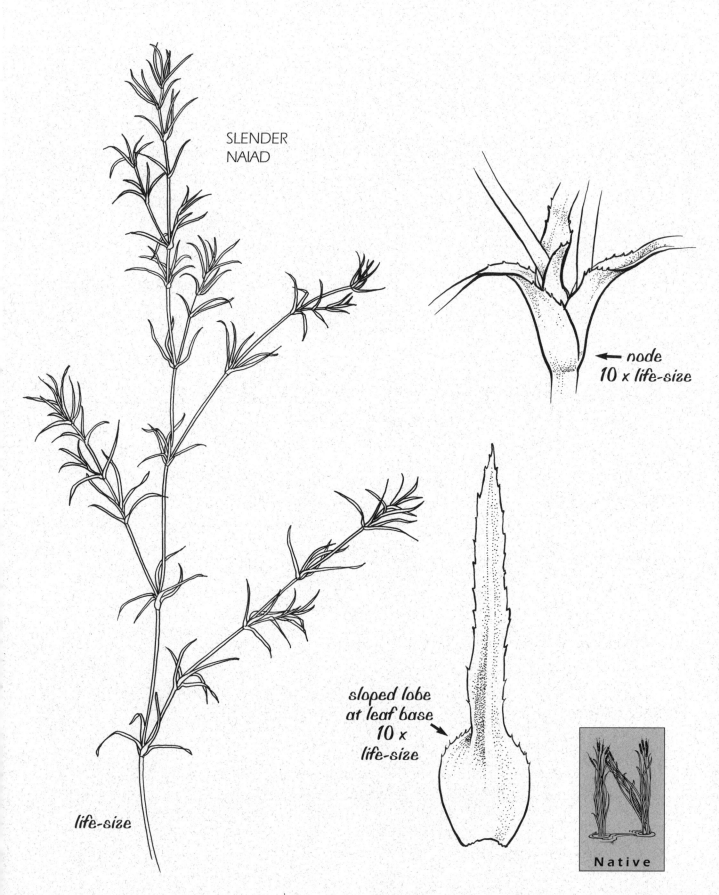

SLENDER
NAIAD

← node
10 x life-size

sloped lobe
at leaf base
10 x
life-size

life-size

Native

Najas flexilis (continued)

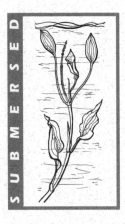

Spiny naiad (*Najas marina*) has larger leaves (0.5-4 cm long, 0.5-4.5 mm wide) that are coarsely toothed. The backs of the leaves have spines along the midvein. The reddish-brown seed (2.2-4.5 mm) has an irregularly pitted surface. Spiny naiad is usually found in very alkaline lakes or ponds and can grow aggressively.

Origin & Range: Native; common throughout Wisconsin; range includes northern and central U.S.

Habitat: Slender naiad will grow at a wide range of depths from very shallow to several meters deep. It is often found in sand or gravel sediment growing in association with wild celery (*Vallisneria americana*).

Through the Year: Slender naiad is a true annual. It dies back completely in the fall and relies on its seeds to return in the spring. Seeds germinate in spring and leafy stems are evident by early summer. Slender naiad often spreads by stem fragments during the growing season. Flowering occurs by midsummer and seeds are evident by mid- to late summer.

Value in the Aquatic Community: Slender naiad is one of our most important plants for waterfowl. Stems, leaves and seeds are all consumed by a wide variety of ducks including black duck, bufflehead, canvasback, gadwall, mallard, pintail, redhead, ringnecked duck, scaup, shoveler, blue-winged teal, green-winged teal, wigeon and wood duck. It is also important to a variety of marsh birds as well as muskrats. Slender naiad is a good producer of food and shelter for fish.

A Closer Look:

Mythical naiads were said to be immortal. Although usually gentle, they could take revenge on people who damaged things under their protection. The plants known as naiads share some of these qualities. They are successful survivors, giving them a seeming immortality. You could say they take revenge on people when water is mistreated; in disturbed conditions, *Najas* can grow at nuisance levels.

Being an annual has advantages and disadvantages for slender naiad. Seeds are more durable and mobile than rhizomes or rootstalks; they can drift into openings or survive in the sediment. They also provide genetic diversity that can be a big advantage in changing conditions. Drawbacks include the energy required to produce seeds, and the smaller energy reserves in seeds versus rhizomes or tubers.

Similar Species:
SPINY NAIAD

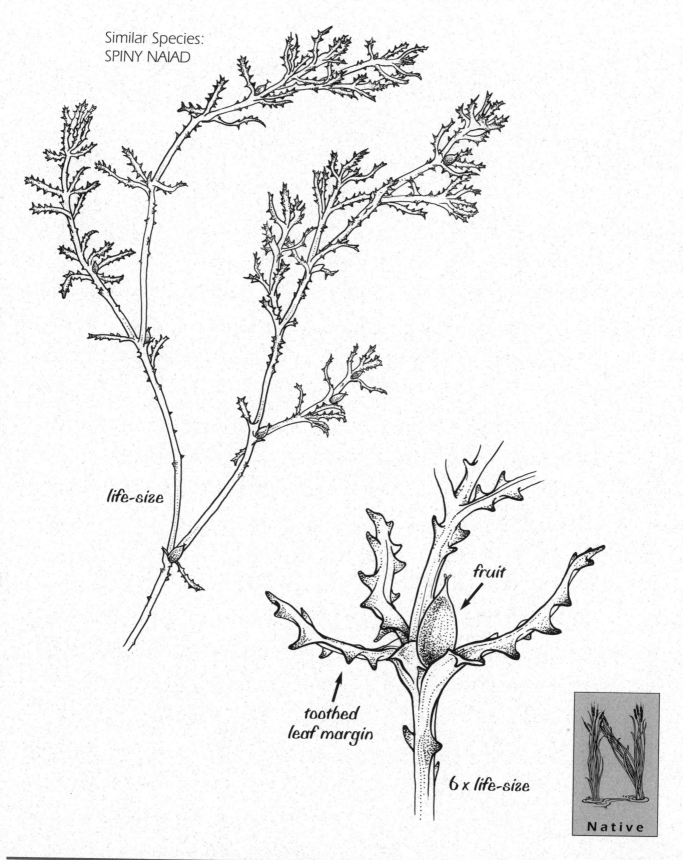

life-size

fruit

toothed
leaf margin

6 x life-size

Native

Nitella spp. (ni–TELL–a)

Nitellas, stoneworts

Nitella – (L.) *nitere:* to shine, *ella:* little one

*Fishing had been good near the drop-off. As she pulled up
the anchor, the woman noticed strands of a branched, green
plant that looked like gelatin . . . Nitella.*

Description: *Nitella* is a type of algae that looks like a higher plant. It has no conductive tissue and has simple anchoring structures called rhizoids rather than true roots. Branches are arranged in whorls around the stem. Stems and branches are smooth and translucent green. The overall plant ranges in size from about 10 cm to 0.5 meter. *Nitella* can reproduce vegetatively by spreading rhizoids as well as sexually. The male reproductive structure, called an antheridium, and female reproductive structure, called an oogonium, are located at the base of the branches. Each pear-shaped oogonium is capped with ten cells.

Similar species: *Nitella* is similar in appearance to muskgrass. However, the stems and branches are smooth rather than lined and encrusted like those of muskgrass. An inspection of the oogonia can also be used to identify the plant. The oogonia of muskgrass have a cap of five cells, while those of nitella have ten cells. *Nitella* also lacks the skunky smell that muskgrass possesses.

Origin & Range: Native; common throughout Wisconsin; range includes most of U.S.

Habitat: *Nitella* is often found on soft sediments in the deeper zones of lakes, sometimes in water depths of 10 meters or more.

Through the Year: *Nitella* overwinters by rhizoids and fragments. Growth begins when the water warms in spring and continues through the fall.

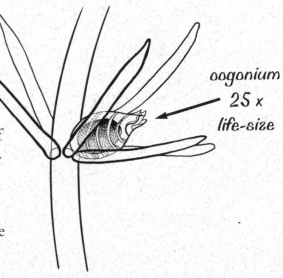

*oogonium
25 x
life-size*

Value in the Aquatic Community:

Nitella is sometimes grazed by waterfowl. The algae and invertebrates on
its surface are attractive to ducks and
geese. *Nitella* also offers foraging opportunities for fish.

A Closer Look:

Take a look at *Nitella* with a magnifying
glass. You'll see that a vast array of micro-
scopic algae, called diatoms and desmids,
as well as tiny invertebrates cling to the
branches. Look espe-
cially for dark-spotted
haliplid beetles (5 mm
long) that make their
home on *Chara* and
Nitella. These crawling
water beetles feed on
algae and tiny animals
that coat the stems and
branches.

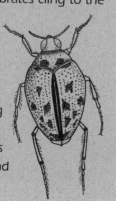

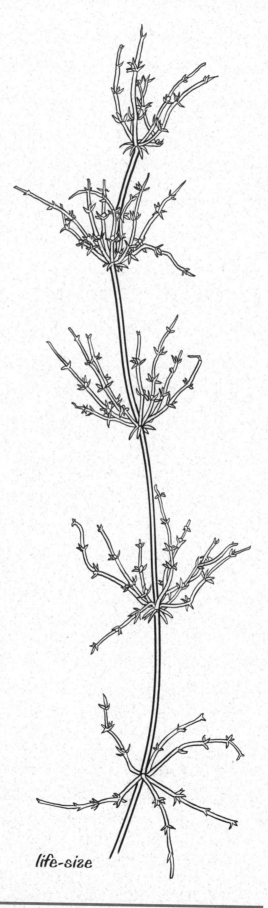

life-size

Native

Ruppia maritima (RUP-ee-a MAR-i-time-a)

Ditch-grass, wigeon-grass

Ruppia – named for Heinrich B. Ruppius, a German botanist (1689-1719);
maritima – (L.) of the sea

The storm had lasted for two days. Only the wigeon and canvasbacks were still on the agitated water. Attracted by the ditch-grass fruit, they dove down to scoop up the nutlets. Many of the shallow-rooted plants had washed onshore in long windrows.

Description: The trailing stems of ditch-grass emerge from a shallow root system. Stiff, slender leaves (3-10 cm long, 0.5 mm wide) are scattered on the stems. Each leaf has an expanded, open sheath at the base. The flower stalk is long and spirally twisted. As the flowers mature, the fruit is elevated on stalks in an umbrella-like cluster called an umbel.

Similar species: Ditch-grass could be confused with a narrow-leaf pond-weed or possibly water bulrush (*Scirpus subterminalis*). Look for the expanded, open sheath at the leaf base. When fruit is present, identification is easy – the umbel arrangement of ditch-grass is very different than the spikes of pondweed or spikelets of water bulrush.

Origin & Range: Native; a few scattered locations in southeastern Wisconsin; range includes coastal areas of the U.S. as well as some inland sites.

Habitat: Ditch-grass is usually found in brackish, saline, or very alkaline water. It can be found growing in water several meters deep.

Through the Year: Ditch-grass can overwinter by seed or rhizome. New shoots develop in early summer. Flowers develop by midsummer, followed by elongation of the fruiting stalk and maturation of the fruits.

Value in the Aquatic Community: Ditch-grass is popular with waterfowl including black duck, bufflehead, canvas-back, gadwall, goldeneye, mallard, pintail, redhead, ringneck, scaup, shoveler and wigeon. They graze on both foliage and fruit. Ditch-grass is also considered an excellent source of food and cover for fish.

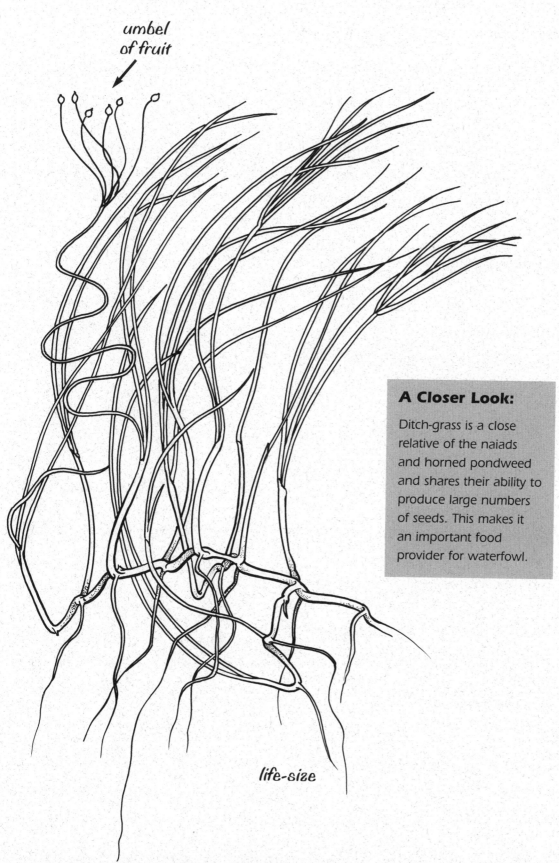

umbel
of fruit

life-size

A Closer Look:

Ditch-grass is a close relative of the naiads and horned pondweed and shares their ability to produce large numbers of seeds. This makes it an important food provider for waterfowl.

Native

SUBMERSED

Zannichellia palustris (zan-i-KEL-ee-a pa-LUS-tris)

Horned pondweed

Zannichellia – named for Gian Zannichelli, an Italian botanist (1662-1729);
palustris – (L.) of marshes

*In less than a decade the land around the little lake had gone
from corn fields to condos. It used to be a favorite place for families
to boat. Horned pondweed was common there. The leaves and
stems were as fine as fish line. The slender plant often went
unnoticed intertwined with other more robust plants.*

Description: Horned pondweed has slender stems that emerge from an equally slight rhizome. The long, narrow leaves (3-10 cm long, 0.5 mm wide) are opposite, although they sometimes appear whorled near the ends of the stems. Flowers develop in the leaf axils and produce a flattened fruit (2-3 mm long) that is slightly curved with a wavy margin and persistent beak (1 mm long).

Similar species: Horned pondweed could be confused with fine-leaved pondweeds or naiads. It can be separated from *Potamogeton* spp. by its opposite rather than alternate leaves. It differs from *Najas* spp. in having longer, less crowded leaves. The best distinguishing feature is the fruit, which is very different from both the pondweeds and naiads.

Origin & Range: Native; scattered locations in Wisconsin, primarily in the southern part of the state; range includes most of the U.S.

Habitat: Horned pondweed is often partly buried in silt or mud, making it difficult to spot. It can be found from shallow zones to water several meters deep.

Through the Year: Horned pondweed is an annual that relies on its seeds to overwinter. Seeds germinate in spring; rhizomes and stems are evident by early summer. Flowers develop in leaf axils by midsummer and the flat fruit can be found later in the summer.

Value in the Aquatic Community: The fruit and foliage of horned pondweed are grazed by waterfowl including black duck, gadwall, mallard, northern pintail, redhead, ring-necked duck, scaup and shoveler. It is also considered a fair food producer for trout.

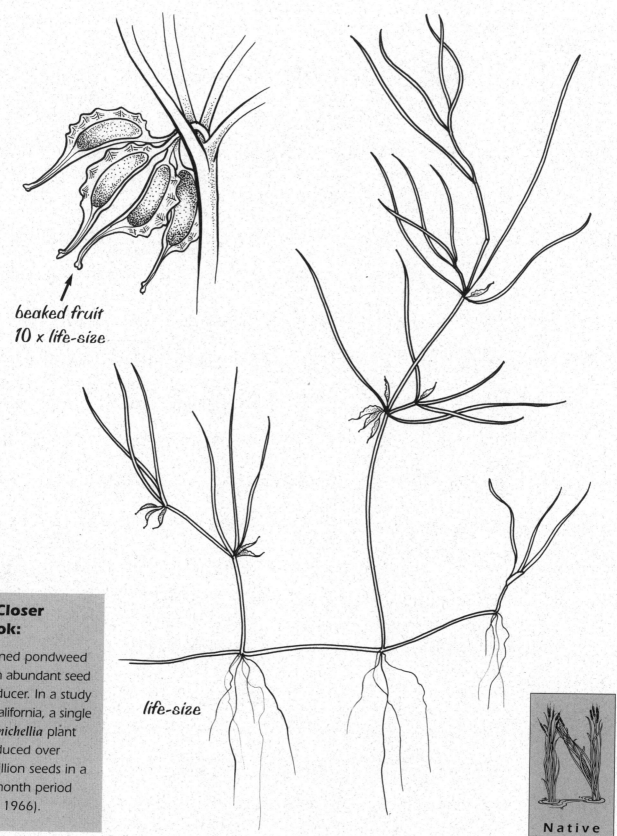

beaked fruit
10 x life-size

life-size

A Closer Look:

Horned pondweed is an abundant seed producer. In a study in California, a single *Zannichellia* plant produced over 2 million seeds in a six-month period (Yeo 1966).

Native

SUBMERSED

Eriocaulon aquaticum (er-ee-oh-CALL-on ah-KWA-ti-cum)

(formerly known as *Eriocaulon septangulare*)

Pipewort

Eriocaulon – (Gk.) *erion:* wool + *kaulos:* stalk; *aquaticum* – (L.) of the water

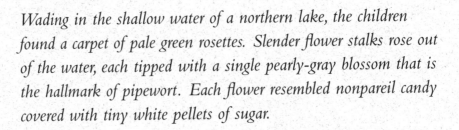

Wading in the shallow water of a northern lake, the children found a carpet of pale green rosettes. Slender flower stalks rose out of the water, each tipped with a single pearly-gray blossom that is the hallmark of pipewort. Each flower resembled nonpareil candy covered with tiny white pellets of sugar.

Description: Pipewort has pale unbranched roots with closely-spaced partitions that make them look cross-hatched. The translucent green leaves (2-5 mm wide, 2-10 cm long) grow in a basal rosette. Leaves taper from base to tip and have a checkerboard appearance created by many short cross-veins.

Each rosette usually produces a single flower stalk. The stalk is slightly twisted with 5-7 ridges. It can range from a few centimeters to a couple meters in length depending on the depth of the water. The flower head (4-6 mm) is rounded with many small flowers packed closely together. Sepals and petals of the pearl-colored flowers are tipped with fine white hairs.

Similar species: Pipewort may be confused with other small rosette forming species such as quillwort (*Isoetes* sp.) or water lobelia (*Lobelia dortmanna*). However, the cross-hatched roots and button-like flowers distinguish pipewort.

Origin & Range: Native; common in soft-water lakes of central and northern Wisconsin; range includes eastern and central U.S.

Habitat: Pipewort is usually found on sandy shores and in shallow water of soft-water lakes. It needs good water clarity and will grow from moist shore-lines to water over 2 meters deep.

Through the Year: The rosettes of pipewort can overwinter green in some circumstances, but on exposed sites it freezes and must grow back from the perennial rootstalk. Flowers are produced by early to midsummer. Small capsular fruits develop by late summer.

Value in the Aquatic Community: Beds of pipewort create shallow water structure for young fish, amphibians and invertebrates. The leaves are sometimes grazed by ducks including black duck and American wigeon.

basal leaf
2 x life-size

A Closer Look:

It's not unusual to see a hummingbird moth or bee hovering over a pipewort blossom. That's because the petals have a small nectar gland near their tip. While many aquatic plants rely on wind or water for cross-pollination, pipewort attracts insects to serve as pollinators.

segmented root

life-size

Native

Isoetes spp. (ice-OH-et-eez)

Quillworts

Isoetes – (Gk.) *isos:* equal + *etos:* year (probably referring to evergreen nature of the leaves)

The clear water and sandy bottom made this the perfect place to swim on a sweltering afternoon. Looking down from the surface, the compact rosettes of the quillwort, tucked into the sediments, looked like bedding plants freshly set out in a garden.

Description: The leaves of quillworts grow out of a fleshy, lobed, underground stem that has forked roots. Each leaf has a central vein and four longitudinal air chambers that can be seen in cross section. Spores form in sacks located on the spoon-like bases of the leaves. Two types of spores are produced on different leaves: megaspores, about the size of sugar crystals, and microspores, as fine as baking powder.

Similar Species:
There are two species of *Isoetes* found in this region:

Spiny-spored quillwort (*Isoetes echinospora,* also known as *I. braunii*) has soft, pale to medium green leaves (5-15 cm long, 0.5-1.5 mm wide) that gradually taper from their base to a long, slender tip. The spore sacs (4-7 mm) are often brown-spotted when fully developed. Mature megaspores are needed for species iden-

SPINY-SPORED
QUILLWORT

*spiny
megaspore
60 x life-size*

LAKE QUILLWORT

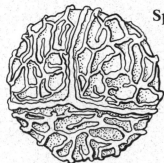

*ridged megaspore
60 x life-size*

*spore
sac
5 x
life-size*

tification. The megaspores (0.25-0.6 mm wide) of spiny-spored quillwort are covered with short spines.

Lake quillwort (*Isoetes lacustris,* also known as *I. macrospora*) has leaves (5-10 cm long, 0.7-2 mm wide) that are dark green, firm and often twisted. The spore sacs (3-5 mm) are pale and usually not spotted. The mature megaspores (0.5-0.8 mm wide) have a convoluted network of ridges on their surface.

A hybrid of *I. lacustris* and *I. echinospora* is sometimes found and has been named *Isoetes* x *hickeyi*. It exhibits a blend of features from the two species. Other species that may be confused with quillwort include plantain shoreweed (*Littorella uniflora*) and pipewort (*Eriocaulon aquaticum*). The leaves of plantain shoreweed can look similar, but the rosettes are connected by rhizomes and the leaves have two air chambers rather than four. Pipewort also has tapered leaves, but the cross-hatched roots and button-like flowers set it apart.

Origin & Range: Native; occasional in soft-water lakes of northern and central Wisconsin; range includes northern and western portions of U.S.

Habitat: Spiny-spored quillwort grows on soft or sandy sediment in water depths ranging from a few centimeters to several meters. Lake quillwort is usually found in sand or gravel and often in water 1-3 meters or more deep. Both species show a preference for soft-water lakes, ponds or streams.

Through the Year: Quillworts overwinter with dark green leaves, or the leaves are lost. In either case, bright green leaves are produced in spring. Spores develop by midsummer and are released when the spore sac decays at the end of the growing season. Spores may germinate near the parent plant or be carried to new sites by waves and currents. Viable spores have also been found in worm feces, so other organisms may aide in dispersal (Boston and Adams 1987).

Value in the Aquatic Community: Quillworts provide habitat in low nutrient lakes that may have very limited plant growth. The foliage is sometimes consumed by waterfowl or game birds including sharp-tailed grouse.

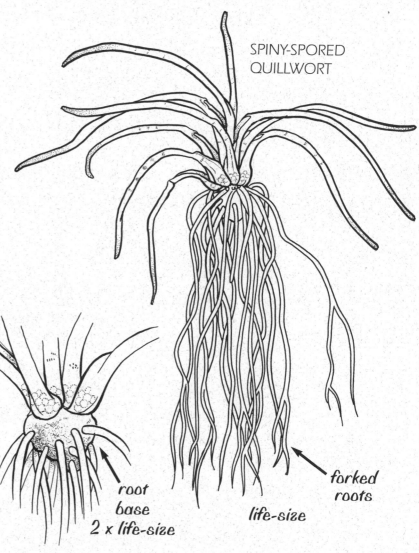

SPINY-SPORED QUILLWORT

root base
2 x life-size

forked roots

life-size

A Closer Look:

Isoetes is the name-sake for a group of plants known as the "isoetids." Some of the plants in this group include quillwort, pipewort (*Eriocaulon aquaticum*), plantain shoreweed (*Littorella uniflora*), water lobelia (*Lobelia dortmanna*) and golden pert (*Gratiola aurea*). These plants are compact, slow-growing evergreens with special adaptations for successful growth in carbon-poor, low-nutrient habitat. Among the adaptations are small size, low leaf turnover, high root-to-shoot ratio, and slow growth rate (Madsen 1991).

Native

Littorella uniflora (LIT-or-el-a U-ni-flor-a)

Plantain shoreweed, littorella

Littorella – (L.) *littor:* seashore + *ella:* little one; *uniflora* – (L.) *uni:* one + *flora:* flower

In the clear water of a northern lake the compact, green tufts of plantain shoreweed had gained a foothold, poking out of the sand as stiff as porcupine quills. Their only visitor was the occasional blue heron that walked among them while searching for a meal.

Description: Plantain shoreweed has a rosette of dark green leaves (4-5 cm long) that emerge from a well-developed rootstalk. The leaves are quill-like and taper to a point. Flowers only develop on plants that emerge out of the water. The male flower is held above the leaf cluster on a slender stalk. It has four stamens protruding beyond a small four-lobed blossom. The female flower is urn-shaped and concealed in the base of the leaves.

Similar species: The small green tufts of plantain shoreweed could be confused with creeping spearwort (*Ranunculus flammula*). The arching surface runners (stolons) of spearwort are a good distinguishing feature.

Origin & Range: Native; uncommon in Wisconsin – listed as a **Special Concern** species; range includes some northern portions of U.S.

Habitat: Plantain shoreweed is usually found in soft-water, low pH lakes. It grows on sandy shorelines and in shallow water.

Through the Year: Under favorable conditions, plantain shoreweed may overwinter green. Active growth resumes in spring. New leaves develop from the base during the growing season. Flowering only occurs on plants that emerge from the water or are stranded on shore.

Value in the Aquatic Community: Plantain shoreweed can provide some limited habitat for invertebrates and fish in low fertility lakes.

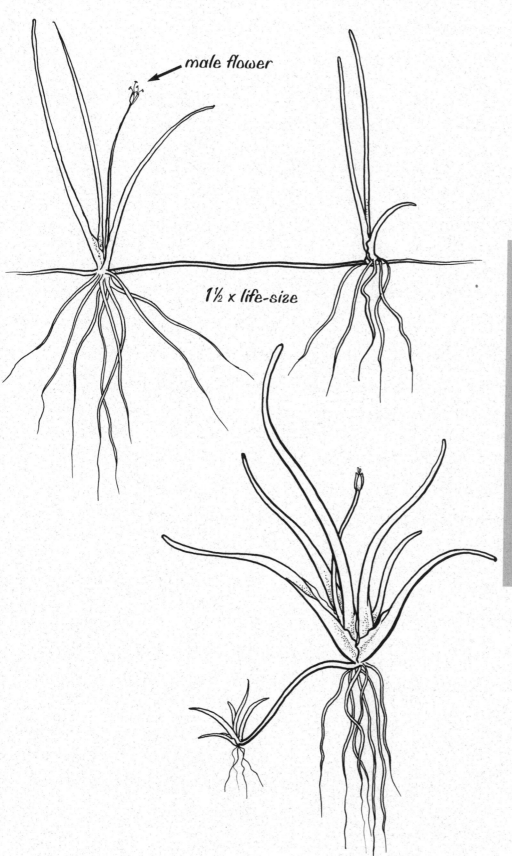

male flower

1½ x life-size

A Closer Look:

If you reach down to touch them, you'll find the leaves of plantain shoreweed are as stiff as the leaves of a plastic aquarium plant. It is an uncommon member of the "isoetid" group – small, evergreen plants that grow in low-nutrient conditions (see discussion at *Isoetes* spp.). *Littorella* is well adapted to grow in lakes that are low in nutrients and dissolved inorganic carbon. It can draw up to 96% of its carbon from the sediment (Madsen 1991).

Rare

Lobelia dortmanna (lo-BEE-lee-a dort-MAN-na)

Water lobelia

Lobelia – named for Matthias de Lobel, a Flemish botanist (1538-1616); *dortmanna* – named for Dortmann, an early Dutch apothecary

The flower-tipped stalk rose above the water from a succulent rosette anchored in the sand. A moth cruised the shore and for a moment hovered over the small, pale blue flowers.

Description: The fleshy, finger-like leaves of water lobelia are arranged in a basal rosette that is anchored by a dense rootstalk. Each leaf (2-9 cm long) is composed of two fused, hollow tubes. The tip of the leaf is rounded and curves outward. Underground stems connect adjacent plants and the roots at each node are white and fibrous.

A single hollow flower stalk (25 cm–1 m tall) emerges from the center of the basal leaves. Plants that flower usually have a minimum of 5-7 leaves. Tubular, pale-blue to white flowers (1-2 cm long) are produced, usually on the portion of the stalk above water level. Like other lobelias, the flowers have two lips. The upper one is divided and the base of the flared lower lip is often hairy.

Similar Species: The fleshy leaves composed of two longitudinal hollow tubes make water lobelia easy to distinguish from other small rosette plants.

Origin & Range: Native; occurs in northern lakes and forest regions of Wisconsin; range includes the eastern U.S. and Pacific coast.

Habitat: Water lobelia is usually found on sandy shores or shallow water (<2.5 m deep) of soft-water lakes.

Through the Year: Water lobelia is a perennial that can resprout from the rootstalk in spring. However, the primary mode of reproduction is by seed. The seeds require a cold period before they can germinate, so seeds sprout in the spring. Plants with well-established rosettes produce a flower stalk and bloom by midsummer. Seeds mature in the capsular fruit by late summer. As the capsule breaks down, seeds are released. The seeds float for a few moments and then sink into the sediment where they overwinter.

Value in the Aquatic Community: Beds of water lobelia can help stabilize sandy, eroding shorelines. It also offers shallow water habitat for invertebrates and young fish.

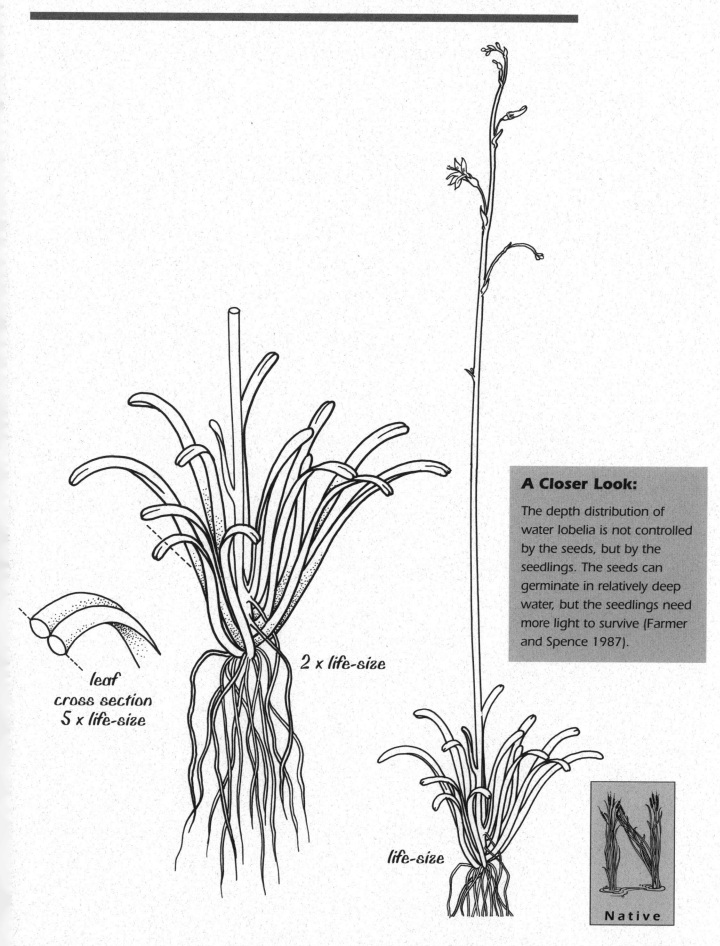

leaf
cross section
5 x life-size

2 x life-size

life-size

A Closer Look:

The depth distribution of water lobelia is not controlled by the seeds, but by the seedlings. The seeds can germinate in relatively deep water, but the seedlings need more light to survive (Farmer and Spence 1987).

Native

SUBMERSED

Myriophyllum tenellum (MIR–ee–o–FIL–um te–NEL–um)

Dwarf water milfoil

Myriophyllum – (Gk.) *myrio*: many + *phyllon*: leaf; *tenellum* – (L.) delicate

The patch of dwarf water milfoil was rarely noticed in the shallow water of the northern lake. It looked like a row of toothpicks neatly stuck in the sand.

Description: Dwarf water milfoil looks very different than other water milfoil species. The slender, unbranched stems (2-15 cm tall) arise singly along a buried rhizome. The leaves are reduced to small scales or bumps. Tips that rise above the water may flower. The pale flowers are small and produce nut-like fruits.

Similar species: The chain of toothpick-like stems gives dwarf water milfoil a unique appearance. Because of its small size, it is often overlooked. Once spotted, it is easy to identify.

Origin & Range: Native; found primarily in northern Wisconsin; range includes northeastern U.S.

Habitat: Dwarf water milfoil occurs primarily on sandy sites out to a depth of about 4 meters. It can form a dense turf of closely spaced stems.

Through the Year: Dwarf water milfoil overwinters as buried rhizomes. New shoots appear in early summer. Plants growing in shallow water may flower in early to midsummer. In some lakes plants may only flower during drought years. The nut-like fruits mature by late summer.

Value in the Aquatic Community: Dwarf water milfoil provides good spawning habitat for panfish and shelter for small invertebrates. The network of rhizomes helps stabilize sediment.

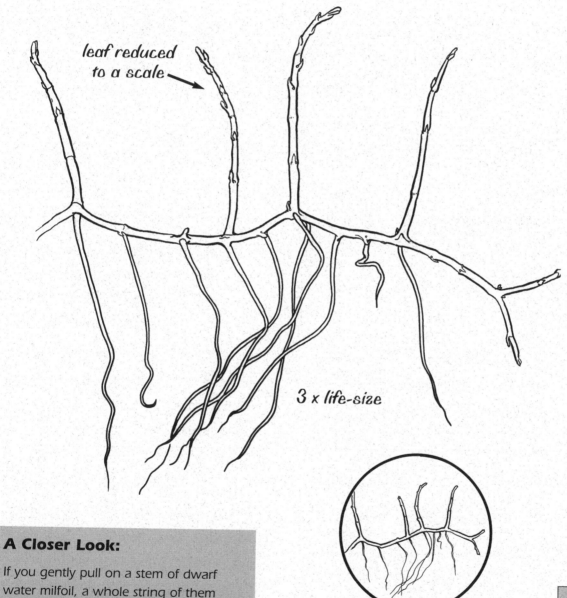

leaf reduced to a scale

3 x life-size

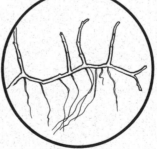

life-size

A Closer Look:

If you gently pull on a stem of dwarf water milfoil, a whole string of them comes out – like the chain stitching on a bag of sugar. The flowers of dwarf water milfoil are wind pollinated. On each individual flower stalk, the female flowers mature before the male flowers. This helps to ensure cross pollination.

Native

Potamogeton spp. (POT-a-mo-JEE-ton)

Pondweeds

Potamogeton – (Gk.) *potamos*: river + *geiton*: neighbor

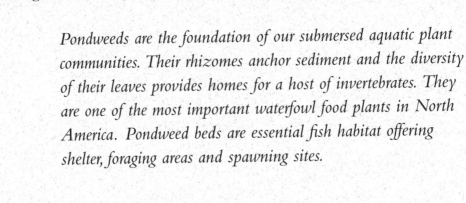

Pondweeds are the foundation of our submersed aquatic plant communities. Their rhizomes anchor sediment and the diversity of their leaves provides homes for a host of invertebrates. They are one of the most important waterfowl food plants in North America. Pondweed beds are essential fish habitat offering shelter, foraging areas and spawning sites.

Description: Identification of pond-weeds is tricky. However, there are some key features about pondweeds that can help you recognize them. They all have alternate, entire leaves with a noticeable midvein. Flowers are small, greenish-brown and arranged in a spike. The fruit is a nutlet called an achene. This achene has distinctive surface features – often very helpful in identification. When the fruits are not present, it is often necessary to take a close look at the leaves and the stipules. The stipule is a bit of tissue associated with the base of each leaf. It may be fibrous or membranous, fused to the leaf or free. If you are willing to take a close look at the leaves and stipules with a hand lens or dissecting scope, you can often determine which species of pondweed you have.

Similar Species: Narrow-leaved pondweeds are sometimes confused with water stargrass (*Zosterella dubia*).

However, the leaves of water stargrass lack a definite midvein. When in bloom, water stargrass is unmistakable because of the yellow, star-shaped flowers.

Origin & Range: All *Potamogeton* spp. in our region are native except curly-leaf pondweed (*Potamogeton crispus*), which was introduced from Europe.

Habitat: Pondweeds grow in a wide range of aquatic habitats from flowing water to still bays, mucky sediment to sand, and shallow water to deep, sub-mersed sites. Some species can live on mud out-of-water during summer droughts.

Through the Year: Most pondweeds overwinter by rhizomes and winter buds. However, some of the broad-leaved species can be "evergreen" under the ice. Seasonal growth patterns vary and some species put more energy into seed production than others.

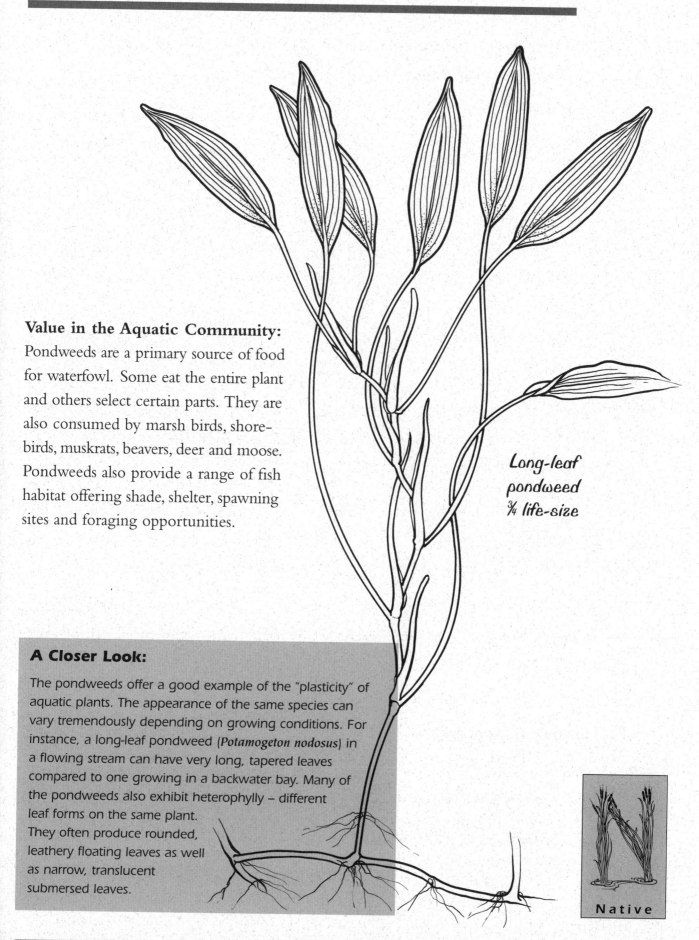

Value in the Aquatic Community:
Pondweeds are a primary source of food for waterfowl. Some eat the entire plant and others select certain parts. They are also consumed by marsh birds, shore–birds, muskrats, beavers, deer and moose. Pondweeds also provide a range of fish habitat offering shade, shelter, spawning sites and foraging opportunities.

Long-leaf
pondweed
¾ life-size

A Closer Look:

The pondweeds offer a good example of the "plasticity" of aquatic plants. The appearance of the same species can vary tremendously depending on growing conditions. For instance, a long-leaf pondweed (*Potamogeton nodosus*) in a flowing stream can have very long, tapered leaves compared to one growing in a backwater bay. Many of the pondweeds also exhibit heterophylly – different leaf forms on the same plant. They often produce rounded, leathery floating leaves as well as narrow, translucent submersed leaves.

Native

Potamageton amplifolius
(POT-a-mo-JEE-ton AM-pli-FOL-ee-us)

Large-leaf pondweed, bass weed, musky weed

Potamogeton – (Gk.) *potamos:* river + *geiton:* neighbor;

amplifolius – (L.) *ampli:* large + *folius:* leaf

The old fishing guide studied the water. At last he saw what he was looking for: the broad arching leaves of large-leaf pondweed rising from the depths. The anchor created a dark cloud as it settled in the soft sediments seven feet below. This was the spot!

Description: Large-leaf pondweed has robust stems (1-3.5 mm thick) that emerge from black-scaled rhizomes (2-4 mm thick). The submersed leaves (3.5-7.2 cm wide) are the broadest of any pondweed in our region. These leaves are arched and slightly folded, as though they had been bent along the midvein to cut out a symmetrical leaf. The leaves have stalks of varying lengths (1-6 cm long) and the blade is lined with many veins (25-37).

Floating leaves are oval (5-10 cm long, 2.5-5 cm wide) on long stalks (8-30 cm).

The floating leaves also have many veins, with about 25% of them more prominent than the rest. Stipules of both submersed and floating leaves are large (3.5-12 cm long). The stipules are free, green or brown-tinted, and taper to a sharp point. (See *Potamogeton* spp. for a description of stipules.)

The fruiting stalk is stout with flowers – and later fruit – in a dense spike (2-5 cm long). Each fruit (4-5.5 mm) is oval to egg-shaped, with a small beak (1 mm long). The surface of the fruit has three low, rounded ridges.

A Closer Look:

Large-leaf pondweed is a premier aquatic plant for fish habitat. The common names bass weed and musky weed recognize this reputation. Stands of large-leaf pondweed are considered ecologically valuable habitat. Efforts have been made to propagate it as part of littoral zone restoration projects. Studies have shown large-leaf pondweed can be successfully transplanted by stem cuttings when conditions are favorable for its growth. In a study on Lac La Belle and Okauchee Lake in Waukesha County, Wisconsin, stem cuttings were planted in late June. These transplants averaged a 10-fold increase in shoot length and extensive rhizome growth during the first growing season (Les 1988).

Similar Species: Large-leaf pond-weed is sometimes confused with Illinois pondweed (*P. illinoensis*) or white-stem pondweed (*P. praelongus*). Illinois pondweed has narrower submersed leaves with fewer veins (9-19). It also has a sharp-keeled fruit and floating leaves with stalks that are shorter than the blades. White-stem pondweed has submersed leaves directly attached to the stem. The stipules are fibrous and the tip of each leaf is usually boat-shaped.

Origin & Range: Native; found throughout Wisconsin; range includes much of U.S.

Habitat: Large-leaf pondweed is found most frequently in soft sediments in water one-to-several meters deep. It is sensitive to increased turbidity and suffers when top-cut by motor boats.

Through the Year: Large-leaf pond-weed can sprout shoots from overwintering rhizomes. It also sometimes survives the winter as an evergreen. Flowering occurs by midsummer and fruits mature by late summer.

Value in the Aquatic Community: The broad leaves of *P. amplifolius* offer shade, shelter and foraging opportunities for fish. Abundant production of large nutlets makes this a valuable waterfowl food.

floating leaf

stipule

¾ life-size

Native

SUBMERSED

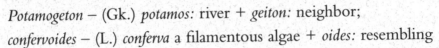

Potamogeton confervoides
(POT-a-mo-JEE-ton CON-fer-VOID-eez)

Algal-leaved pondweed

Potamogeton – (Gk.) *potamos*: river + *geiton*: neighbor;

confervoides – (L.) *conferva* a filamentous algae + *oides*: resembling

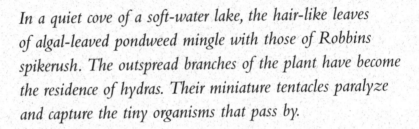

In a quiet cove of a soft-water lake, the hair-like leaves of algal-leaved pondweed mingle with those of Robbins spikerush. The outspread branches of the plant have become the residence of hydras. Their miniature tentacles paralyze and capture the tiny organisms that pass by.

Description: Algal-leaved pondweed has stems that arise from a long, creeping rootstalk. The stems (10-80 cm) are slender and branch repeatedly, creating a fan-shaped appearance. Submersed leaves are very fine (2-5 cm long, 0.25 mm wide) and have only a single vein. No floating leaves are produced. The fruiting stalk is the most distinguishing feature. It extends as a continuation of the main stem and is quite long, ranging from 5-25 cm. The base of the fruiting stalk is slender and gradually thickens toward the compact spike of fruit. Each fruit (2-3 mm long) has three ridges. Tightly-wrapped winter buds (1-2 cm long) often form on side branches.

Similar species: Algal-leaved pondweed is unlikely to be confused with any other species. The combination of fan-shaped branches and a single, stout fruiting stalk is unique.

Origin & Range: Native; algal-leaved pondweed has been found at a few scattered locations in central and northern Wisconsin – listed as **Threatened**; range includes northeastern U.S.

Habitat: Algal-leaved pondweed grows in soft-water, low pH lakes. It is usually found in quiet, shallow water.

Through the Year: Algal-leaved pondweed overwinters by hardy rhizomes. Reproduction from seed may occur if conditions are favorable. New stems sprout as the water warms in spring. The stout fruiting stalk is usually evident by midsummer. Compact winter buds often form late in the growing season.

Value in the Aquatic Community: The foliage and fruit of algal-leaved pondweed may be grazed by a variety of waterfowl. The fan-like branches provide invertebrate habitat and foraging opportunities for fish.

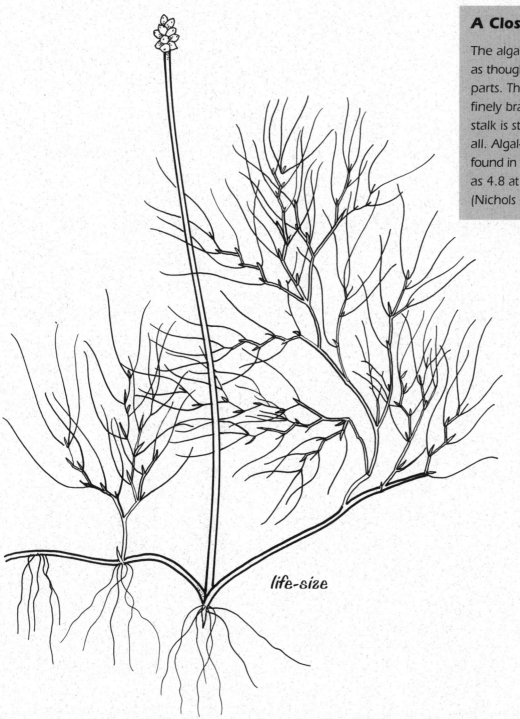

life-size

Rare

Potamogeton crispus (POT-a-mo-JEE-ton CRISP-us)

Curly-leaf pondweed

Potamogeton – (Gk.) *potamos:* river + *geiton:* neighbor; *crispus* – (L.) curly

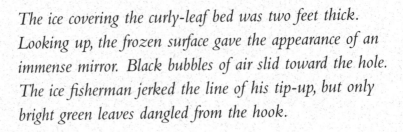

The ice covering the curly-leaf bed was two feet thick. Looking up, the frozen surface gave the appearance of an immense mirror. Black bubbles of air slid toward the hole. The ice fisherman jerked the line of his tip-up, but only bright green leaves dangled from the hook.

Description: The slightly flattened stems of curly-leaf pondweed grow out of a slender rhizome. Although it is a submersed aquatic plant, the spaghetti-like stems often reach the surface by mid-June. Submersed leaves (3-8 cm long, 5-12 mm wide) are oblong and attach directly to the stem in an alternate pattern. Margins of the leaves are wavy and finely serrated, creating an overall leaf texture that is "crispy." The stipules (3-8 mm long) are fused to the base of the leaf and disintegrate as the growing season progresses. (See *Potamogeton* spp. for definition of stipules.) No floating leaves are produced.

In the spring, curly-leaf produces flower spikes that stick up above the water surface. The small flowers are arranged in a terminal spike on a curved stalk (2-5 cm). Fruits develop that each have three ridges and a conical beak (2-2.5 mm long).

serrated leaf margin 3 x life-size

Curly-leaf also produces vegetative buds called turions that look like small, brown pine cones on shortened branches along the stem.

Seeds play a relatively small role in reproduction compared to the turions. Studies of curly-leaf beds in lakes have shown as many as 1,600 turions in just a square meter plot. In some cases, 60-80% of the turions germinate (Nichols 1986). *(continued)*

nutlet 8 x life-size

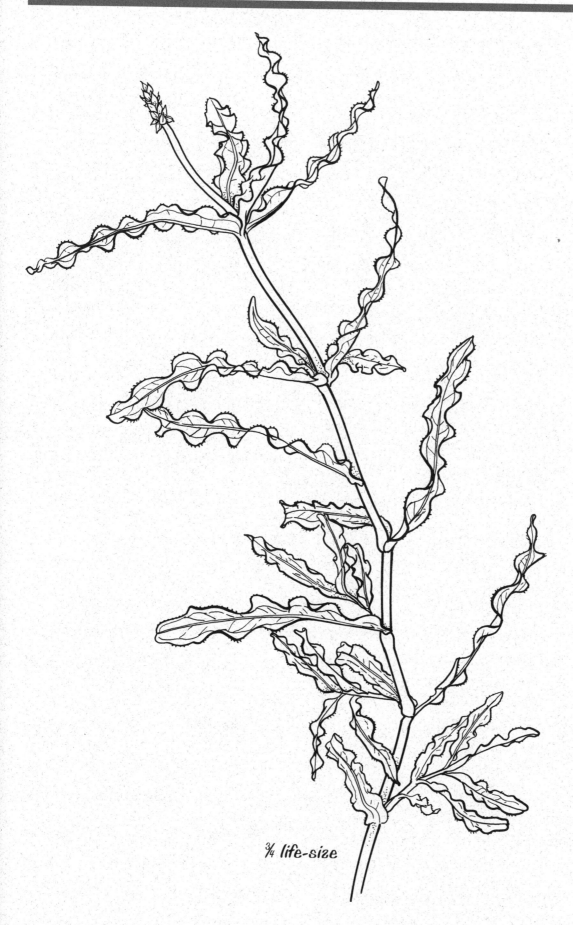

¾ life-size

Potamogeton crispus (continued)

Similar Species: Curly-leaf could be confused with clasping-leaf pondweed (*Potamogeton richardsonii*). The size and shape of the leaves of curly-leaf and clasping-leaf pondweed may be similar, but the leaves of clasping-leaf pondweed don't have finely toothed margins and are heart-shaped at the point where they clasp the stem. Another difference is the lack of pine cone-like turions in clasping-leaf pondweed.

Origin & Range: Exotic. The first confirmed specimen of this European exotic in the U.S. was collected in Delaware in the mid-1800s. The first record of curly-leaf in Wisconsin was in 1905, and it is now common throughout the state. Range includes most of U.S.

Habitat: Curly-leaf pondweed is usually found in soft sediments in water, ranging from less than a meter to several meters deep. It can tolerate low light and will grow in turbid water.

Through the Year: The cool water adaptations of curly-leaf set it apart from other Wisconsin aquatic plants. It grows under the ice while most plants are dormant, but dies back in mid-July when other aquatic plants are just reaching peak growth.

The life cycle of curly-leaf is triggered by changes in water temperature. Warming waters in May stimulate growth of the spring foliage which has wider leaves than the winter growth and wavy leaf margins. During the spring, flowers and fruit may be produced. Then as water temperatures rise in early July, curly-leaf prepares for late summer dormancy. Many turions are produced and the foliage begins to break down. By August, the majority of curly-leaf stems and leaves have decayed and dropped a carpet of sharp-angled turions on the sediment.

These turions lie dormant until the water begins to cool in September.

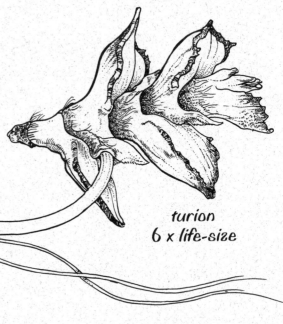

turion
6 x life-size

When the water temperature falls to about 75°F, the turions germinate to produce winter foliage. The winter curly-leaf growth has flat, blue-green leaves that are narrower, softer and more translucent than the summer leaves. The winter growth form of curly-leaf thrives under the ice. It has been found growing under thick ice and a heavy blanket of snow. When the water warms up in May, spring foliage is produced.

Value in the Aquatic Community:
Curly-leaf provides habitat for fish and invertebrates in the winter and spring when most other aquatic plants are reduced to rhizomes and winter buds. However, the midsummer die-off of curly-leaf pondweed creates a sudden loss of habitat and releases nutrients into the water column that can trigger algal blooms and create turbid water conditions.

A Closer Look:

It may seem odd that a plant could be growing under the ice, but curly-leaf pondweed is a cool-water specialist. This exotic species got used to cold temperatures in its home range of European and Asian streams. The split-season growth cycle of curly-leaf pondweed makes this plant unique in the Wisconsin aquascape. While it provides winter habitat for fish and invertebrates, its rapid growth in the spring can create problems for recreation and navigation. The midsummer die-off can also degrade water quality. Selective control of curly-leaf is often needed. Protection or restoration of native species can lead to a balanced plant community. Protecting or improving water quality can also help keep curly-leaf in check because it has a competitive advantage over native plants when water clarity is reduced.

Exotic

Potamogeton diversifolius
(POT-a-mo-JEE-ton DIE-ver-si-FOL-ee-us)

Water-thread pondweed, variable-leaf pondweed, snailseed pondweed

Potamogeton – (Gk.) *potamos*: river + *geiton*: neighbor;
diversifolius – (L.) *diversi*: different + *folius*: leaves

*Water-thread rises from the sediments in a silent green underwater
thicket. A snail grazes on the narrow leaves at its trademark pace
as a school of perch lingers in the emerald shadows.*

Similar Species:
VASEY'S
PONDWEED

½
life-size

nutlet
8 x life-size

Description: Water-thread pondweed
has freely-branched stems that emerge
from a slender, buried rhizome. Submersed leaves (1-10 cm long, 0.1-1.5 mm
wide) are narrow and linear. There is
one obvious midvein bordered by a
single row of lacunar (hollow) cells. Two
faint lateral veins may also be visible.
The stipules are fused to the leaf for
one-third to one-half of their length,
creating a tongue-like appendage at
the free end. (See *Potamogeton* spp. for
definition of stipules.) Floating leaves
(0.5-4 cm long, 2-20 mm wide) are
shaped like an ellipse.

Fruiting spikes may be present in two
forms. Those formed in the axils of
submersed leaves are globe-shaped with
a short stalk (1-10 mm) and 1-15 fruits.
Those produced in the axils of floating
leaves have longer stalks (3-32 mm long)
with a cylindrical spike of 5-120 fruits.
The fruit (1-2 mm) is round and flat
with a tiny beak, a raised dorsal ridge
and two lower lateral ridges that may
appear as teeth. The embryo coil can

be seen through the wall of the fruit,
giving it a snail-like appearance.

Similar species: It takes a sharp eye
to recognize *Potamogeton diversifolius*
because it can appear in different forms.
It could be confused with spiral-fruited pondweed (*Potamogeton spirillus*)
which has submersed leaves that are
usually curved and slightly wider than
water-thread pondweed. The stipule
of spiral-fruited pondweed is fused to
the leaf for most of its length, creating
only a free tab rather than the longer
tongue visible on water-thread. Fruits
of spiral-fruited also have the snail-like
appearance, but there is only a sharp
dorsal ridge with no side ridges (see
discussion at *Potamogeton spirillus*).

Water-thread pondweed may also be
confused with **Vasey's pondweed**
(*Potamogeton vaseyi*), which has fine,
hair-like leaves (2-6 cm long, 0.2-1
mm wide) similar to the more slender-
leaved varieties of water-thread.
Stipules are completely free from
the leaves. Floating leaves (8-15 mm

long) are shaped like an ellipse and have slender, hollow cells between the veins. The fruit (1.2-2.5 mm long) is rounded without the snail-like appearance of water-thread or spiral-fruited pondweed. Numerous winter buds are often produced on side branches. Vasey's pondweed is listed as a **Special Concern** species in Wisconsin.

Some botanists recognize two varieties of water-thread pondweed.

P. diversifolius **var.** *diversifolius* (also known as *P. capillaceus*) has long, fine leaves that are usually about 0.5 mm wide. *Potamogeton capillaceus* is listed as a **Special Concern** species in Wisconsin.

P. diversifolius **var.** *trichophyllus* (also known as *P. bicupulatus*) has filament-like leaves that are only about 0.2 mm wide.

Origin & Range: Native; infrequent to rare in Wisconsin; range includes portions of U.S. though more common in the south.

Habitat: Water-thread pondweed can be found in lakes, ponds, rivers and streams. It is usually found on soft sediment in water less than 2 meters deep.

Through the Year: New shoots are produced in spring from overwintering rhizomes. Flowering occurs by midsummer and fruit is evident by late summer.

Value in Aquatic Community: The fruit produced by water-thread pondweed can be a locally important food source for a variety of ducks and geese. The plant may also be grazed by muskrat, deer, beaver and moose. Leaves and stems may be colonized by invertebrates and offer foraging opportunities for fish.

WATER-THREAD PONDWEED

life-size

Native

A Closer Look:

Another common name for *Potamogeton diversifolius* is snailseed pondweed, in recognition of the snail-like shape of the fruit. Studies have shown northern pintail like to graze on these crunchy fruits.

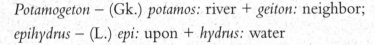

Potamogeton epihydrus (POT-a-mo-JEE-ton EP-i-HIGH-drus)

Ribbon-leaf pondweed

Potamogeton – (Gk.) *potamos:* river + *geiton:* neighbor;
epihydrus – (L.) *epi:* upon + *hydrus:* water

The channel was a major link in the chain of lakes. Luxuriant ribbons of striped foliage draped in the water. They undulated gently in the wake of the boat passing by.

Description: Ribbon-leaf pondweed has slightly flattened stems (up to 2 m long) that emerge from spreading rhizomes. Long, tape-like leaves (up to 20 cm long, 2-10 mm wide) attach directly to the stem (no leaf stalk). The leaves have parallel sides, 5-13 veins, and a prominent stripe of pale green, hollow cells along each side of the midvein. The stipules (1-3 cm long) are not fused to the leaf. (See *Potamogeton* spp. for definition of stipules.) Floating leaves (3-7 cm long, 8-20 mm wide) are supported by a slender leaf stalk about as long as the blade. The floating leaves have the outline of an egg or ellipse.

Fruiting stalks (2-5 cm) are about as thick as the stem. The cylindrical spikes (1-3 cm) are packed with fruit. Each fruit looks like a flattened disk (2.5-4 mm) with slightly indented sides, a dorsal ridge and two lower lateral ridges.

Similar species: Ribbon-leaf pondweed has a unique appearance among the pondweeds of our region. Viewed from the surface, the striped submersed leaves can resemble wild celery (*Vallisneria americana*). However, as soon as you pull them out of the water you can see the leaves are alternate on the stem rather than in a basal rosette like wild celery.

Origin & Range: Native; common in northern Wisconsin; range includes most of U.S.

Habitat: Ribbon-leaf pondweed is found in lakes, ponds and streams, particularly those with low alkalinity. It grows in a variety of sediments in water ranging from knee deep to 1-2 meters.

Through the Year: Stems emerge in spring from overwintering rhizomes and tubers. Flowers appear by early to midsummer and fruits develop by mid-growing season.

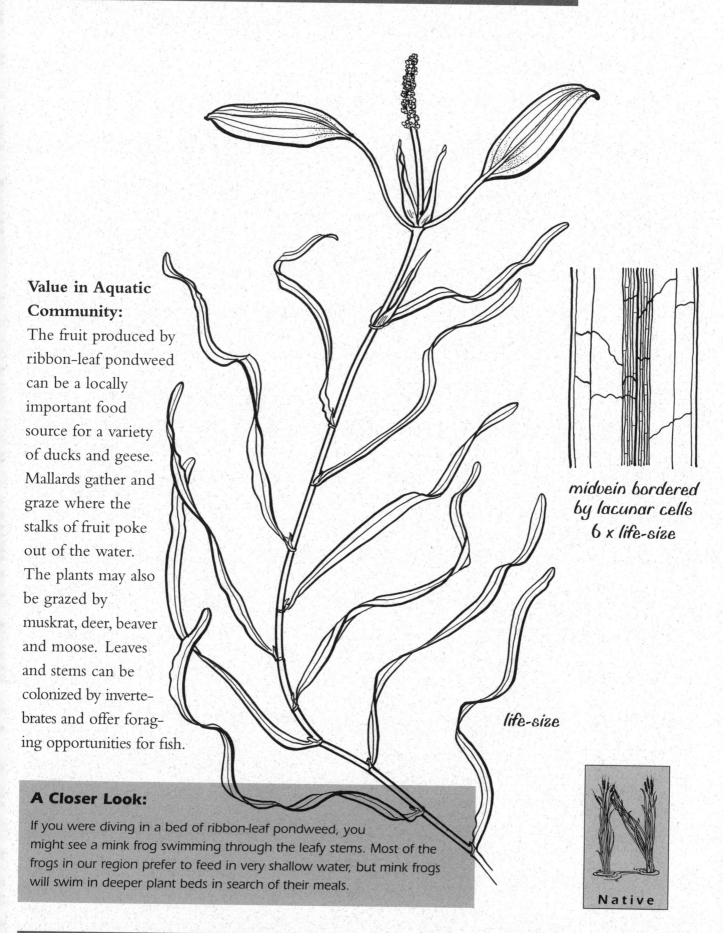

Value in Aquatic Community:

The fruit produced by ribbon-leaf pondweed can be a locally important food source for a variety of ducks and geese. Mallards gather and graze where the stalks of fruit poke out of the water. The plants may also be grazed by muskrat, deer, beaver and moose. Leaves and stems can be colonized by invertebrates and offer foraging opportunities for fish.

midvein bordered by lacunar cells 6 x life-size

life-size

A Closer Look:

If you were diving in a bed of ribbon-leaf pondweed, you might see a mink frog swimming through the leafy stems. Most of the frogs in our region prefer to feed in very shallow water, but mink frogs will swim in deeper plant beds in search of their meals.

Native

Potamogeton foliosus (POT-a-mo-JEE-ton FO-lee-OH-sus)

Leafy pondweed

Potamogeton – (Gk.) *potamos*: river + *geiton*: neighbor; *foliosus* – (L.) leafy

The outfall pipe drained from the storm sewer into the cloudy waters of the bay. The leafy pondweed that grew there was an accommodating plant. Its beds were bustling with small fish, and geese gathered to nibble its fruit.

Description: Leafy pondweed has freely branched stems that emerge from slender rhizomes. The narrow, submersed leaves (1.5-8 cm long, 0.5-2 mm wide) have parallel sides that narrow slightly where they attach to the stem. The tip of the leaf usually tapers to a point. There are 3-5 veins, with the midvein sometimes flanked by 1-2 rows of lacunar cells. The membranous stipules are free from the leaves, but when they are young they wrap around the stem. (See *Potamogeton* spp. for definition of stipules.) No floating leaves are produced.

Flowers and fruit are produced on short stalks (5-15 mm) in the axils of upper leaves. The fruits (4-10) are in tight clusters about 4 mm in diameter. Each fruit is flattened with a dorsal, wavy ridge and a short beak (0.2-0.6 mm). Winter buds are sometimes produced for vegetative reproduction. These buds are composed of several modified, tightly rolled leaves on the ends of branches.

stipule
8 x
life-size

Similar Species: When flowers or fruit are not present, leafy pondweed could be confused with small pondweed (*P. pusillus*). However, small pondweed has glands at the leaf nodes. Fortunately, leafy pondweed blooms early in the season and the short flower stalk with a tight cluster of flowers distinguishes it from small pondweed that has a longer, slender stalk with spaced whorls of flowers.

Origin & Range: Native; scattered locations in Wisconsin; range includes most of U.S.

Habitat: Leafy pondweed can grow in a wide variety of habitats, from backwaters of the Mississippi River to wastewater treatment ponds. Leafy pondweed is most often found in shallow water, but can sometimes be found in water 1 meter or more deep. It shows a preference for soft sediments and is tolerant of eutrophic water conditions.

Through the Year: Leafy pondweed overwinters by rhizomes and winter

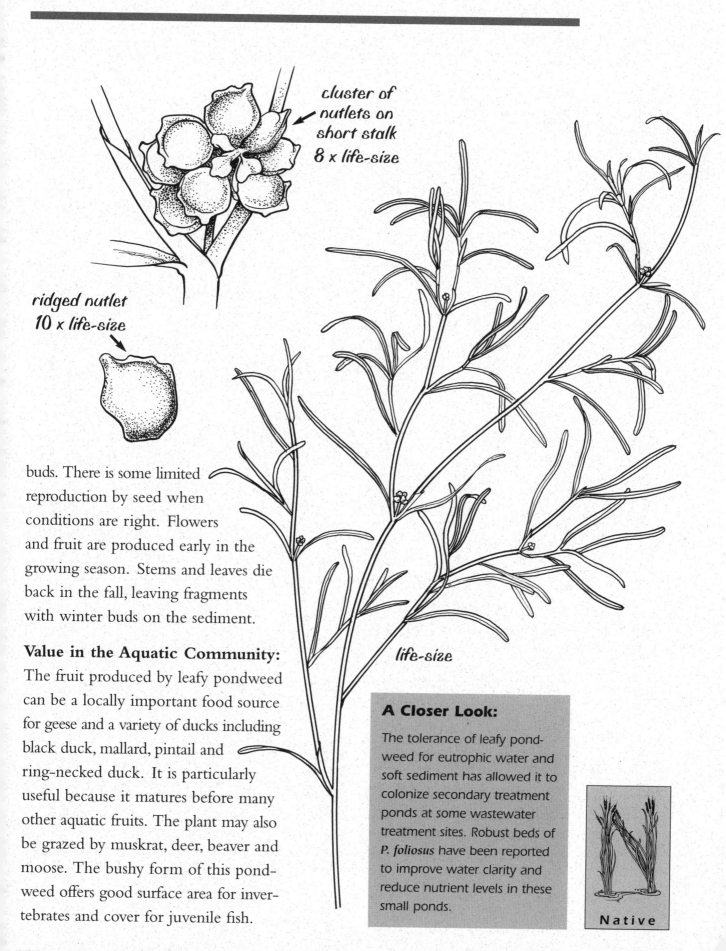

cluster of
nutlets on
short stalk
8 x life-size

ridged nutlet
10 x life-size

life-size

buds. There is some limited reproduction by seed when conditions are right. Flowers and fruit are produced early in the growing season. Stems and leaves die back in the fall, leaving fragments with winter buds on the sediment.

Value in the Aquatic Community:
The fruit produced by leafy pondweed can be a locally important food source for geese and a variety of ducks including black duck, mallard, pintail and ring-necked duck. It is particularly useful because it matures before many other aquatic fruits. The plant may also be grazed by muskrat, deer, beaver and moose. The bushy form of this pond-weed offers good surface area for invertebrates and cover for juvenile fish.

A Closer Look:

The tolerance of leafy pond-weed for eutrophic water and soft sediment has allowed it to colonize secondary treatment ponds at some wastewater treatment sites. Robust beds of *P. foliosus* have been reported to improve water clarity and reduce nutrient levels in these small ponds.

Native

Potamogeton gramineus (POT-a-mo-JEE-ton gra-MIN-ee-us)

Variable pondweed, grass-leaved pondweed

Potamogeton – (Gk.) *potamos:* river + *geiton:* neighbor; *gramineus* – (L.) grasslike

The leech looped through the pondweed bed, finally settling on a lance-shaped leave. Its black shape provided an enticing target for patrolling bass.

Description: Variable pondweed has stems (30-70 cm long, 0.5-1 mm thick) that emerge from spreading rhizomes and often sprawl on the sediment and branch repeatedly. Each side branch has a leafy appearance, with many linear to lance-shaped leaves (3-8 cm long, 3-10 mm wide). The leaves lack stalks, but taper slightly at the point where they attach to the stem. Each leaf has 3-7 veins. Stipules (1-3 cm) are free in the axil of the leaves, with a blunt, slightly hooded tip. (See *Potamogeton* spp. for definition of stipules.) Floating leaves (2-5 cm long, 0.5-2.5 cm wide) have a slender stalk that is usually longer than the blade. The blade is shaped like an ellipse with 11-19 veins.

Flowers and fruit are produced in a dense cylindrical spike (1.5-3 cm long) on a stout flower stalk. The fruit (2-2.5 mm) is flattened with an egg-shaped outline. It has a low dorsal ridge and sometimes faint lateral ridges.

Similar Species: The appearance of variable pondweed can change depending on where it grows – sometimes it is compact with small leaves, other times rangy with larger leaves. It also hybridizes with its broad-leaved neighbors, leading to a blend of pondweed features. It is sometimes difficult to separate larger-leaved versions of *P. gramineus* from Illinois pondweed (*Potamogeton illinoensis*). Usually the extensive branching, smaller stipules and narrower submersed leaves will distinguish variable pondweed from Illinois pondweed.

Origin & Range: Native; common and widely distributed in Wisconsin; range includes northern and western U.S.

Habitat: Variable pondweed is usually found on firm sediment in water about 1 meter deep. However, it can grow in a range of depths from very shallow to several meters deep. It is often found growing in association with muskgrass (*Chara* spp.), slender naiad (*Najas flexilis*) and wild celery (*Vallisneria americana*).

Through the Year: Variable pondweed overwinters by hardy rhizomes and winter buds. Flowering occurs early in the growing season and fruit is produced by midsummer. Foliage usually dies back in late fall.

Value in the Aquatic Community: The fruits and tubers of variable pondweed are grazed by a variety of waterfowl including geese and wood duck. The foliage and fruit may also be eaten by muskrat, beaver, deer and moose. The extensive network of leafy branches offers invertebrate habitat and foraging opportunities for fish.

A Closer Look:

The production of winter buds has been studied in *P. gramineus*. There is considerable variability in the number of winter buds produced each year, but the timing of bud production is quite predictable. Winter buds form when daylight decreases to 10 hours per day (Spencer et al. 1994).

stipule

life-size

Native

Potamogeton illinoensis (POT-a-mo-JEE-ton IL-in-o-EN-sis)

Illinois pondweed

Potamogeton – (Gk.) *potamos:* river + *geiton:* neighbor; *illinoensis* – (L.) of Illinois

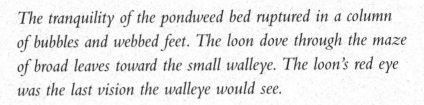

The tranquility of the pondweed bed ruptured in a column of bubbles and webbed feet. The loon dove through the maze of broad leaves toward the small walleye. The loon's red eye was the last vision the walleye would see.

Description: Illinois pondweed has stout stems (up to 2 m long, 1-5 mm wide) that emerge from thick rhizomes. Most of the submerged leaves (8-20 cm long, 2-5 cm wide) are lance-shaped to oval and either attach directly to the stem or have a short stalk (up to 4 cm). These leaves have 9-19 veins and often have a sharp, needle-like tip. The stipules (4-10 cm) are free in the axils of the leaves and have two prominent ridges called keels. (See *Potamogeton* spp. for description of stipules.) Floating leaves which have a thick stalk and ellipse-shaped blade (7-13 cm long, 2-6 cm wide) are sometimes produced. The stalk is usually shorter than the blade.

Flowers and fruits are produced on a stalk (4-12 cm long) that is usually thicker than the stem. The fruit is arranged in a dense cylindrical spike (2.5-6 cm long). Each fruit (3-4 mm wide) has three low dorsal ridges and a short beak (0.5 mm).

Similar Species: Illinois pondweed can be confused with larger-leaved versions of variable pondweed (*P. gramineus*). However, Illinois pondweed has larger stipules with two ridges and broader leaves. Illinois pondweed is also similar in size and shape to **alpine pondweed** (*P. alpinus*). The submersed leaves of alpine pondweed have a central stripe along the midrib, an overall red tint and are more rounded than those of *P. illinoensis*.

Origin & Range: Native; scattered locations in Wisconsin; range includes most of U.S.

Habitat: Illinois pondweed is usually found in water with a moderate to high pH and fairly good clarity. It tends to decline as water becomes turbid. It can grow from shallow zones to water over 3 meters deep.

Through the Year: Illinois pondweed primarily overwinters by winter-hardy rhizomes. However, under some conditions it may overwinter green. Flowering occurs in early summer and fruit is produced by mid-growing season.

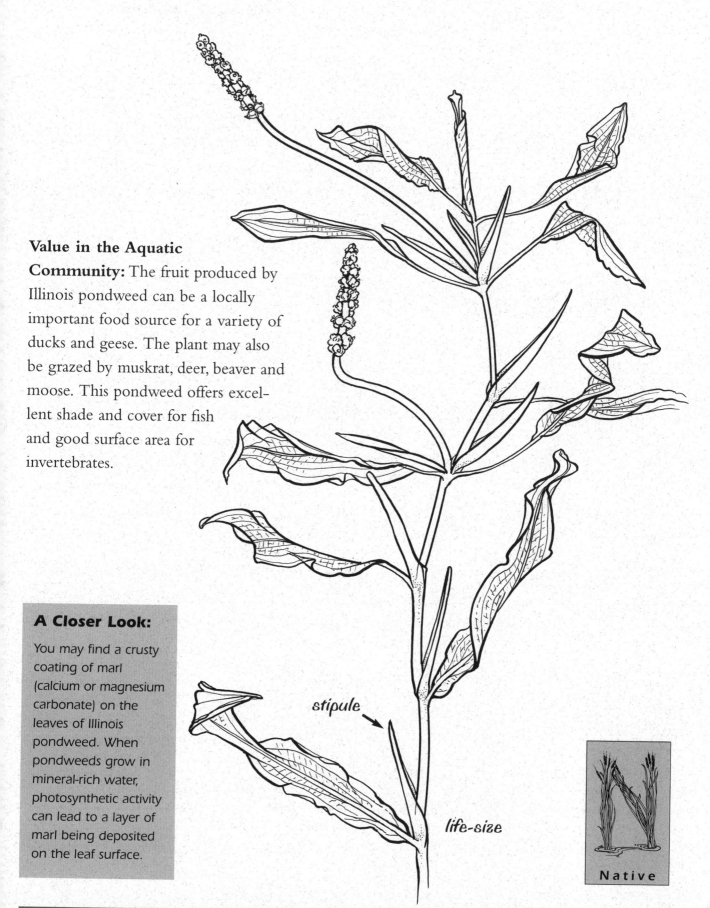

Value in the Aquatic Community: The fruit produced by Illinois pondweed can be a locally important food source for a variety of ducks and geese. The plant may also be grazed by muskrat, deer, beaver and moose. This pondweed offers excellent shade and cover for fish and good surface area for invertebrates.

A Closer Look:

You may find a crusty coating of marl (calcium or magnesium carbonate) on the leaves of Illinois pondweed. When pondweeds grow in mineral-rich water, photosynthetic activity can lead to a layer of marl being deposited on the leaf surface.

stipule

life-size

Native

SUBMERSED

Potamogeton natans (POT-a-mo-JEE-ton NAY-tanz)

Floating-leaf pondweed

Potamogeton – (Gk.) *pòtamos:* river + *geiton:* neighbor; *natans* – (L.) swimming

The floating leaves of the pondweed look like valentines scattered on the water's surface. Each heart-shaped leaf lays flat on the water and rides the waves, anchored by its flexible stalk.

A Closer Look:

When floating-leaf pondweed grows in flowing water, the floating leaves may become elongate and tapered – losing their heart-shaped appearance. This can make it look more like long-leaf pondweed (*P. nodosus*). However, the pale "collar" at the joint of the floating-leaf and stalk is still evident and the submersed leaves are very different than those of long-leaf (Voss 1972).

Description: Floating-leaf pondweed has stems (up to 2 m long) that emerge from red-spotted rhizomes. Submersed leaves (10-40 cm long, 1-2 mm wide) are stalk-like, with no obvious leaf blade. Floating leaves (5-10 cm long, 2-4.5 cm wide) are heart-shaped at their base. The point where the floating leaf attaches to the stalk is distinctive. It looks like someone pinched the stalk and bent it, so the leaf blade is at a right angle to the stalk and lays flat on the water. This "pinched" portion is usually a lighter color than the rest of the stalk. The fibrous stipules of both the submersed and floating leaves are free in the leaf axils. (See *Potamogeton* spp. for a description of stipules.)

Flowers and fruit are produced in a dense cylindrical spike (2-5 cm long) that pokes up above the water surface. Fruit (3.5-5 mm long) is oval to egg-shaped in outline and rather plump. The surface of the fruit has a wrinkled appearance on the sides, a very low dorsal ridge and a short beak.

Similar species: There are two other pondweeds with heart-shaped floating leaves:

Oakes pondweed (*P. oakesianus*) resembles floating-leaf pondweed but it is smaller. Floating leaves are 2.5-6 cm long and submersed leaves are only 1 mm wide. The fruit of Oakes pondweed has a smooth surface compared to the wrinkled fruit of floating-leaf.

Spotted pondweed (*P. pulcher*) has floating leaves that are about the same size as those of floating-leaf pondweed. However, spotted pondweed can be recognized by several key features. The stems and leaf stalks have prominent black spots, the submersed leaves are lance-shaped with a wavy margin and the fruit has three sharp ridges. Spotted pondweed is listed as **Endangered** in Wisconsin.

Origin & Range: Native; common in Wisconsin; range includes northern and western portions of U.S.

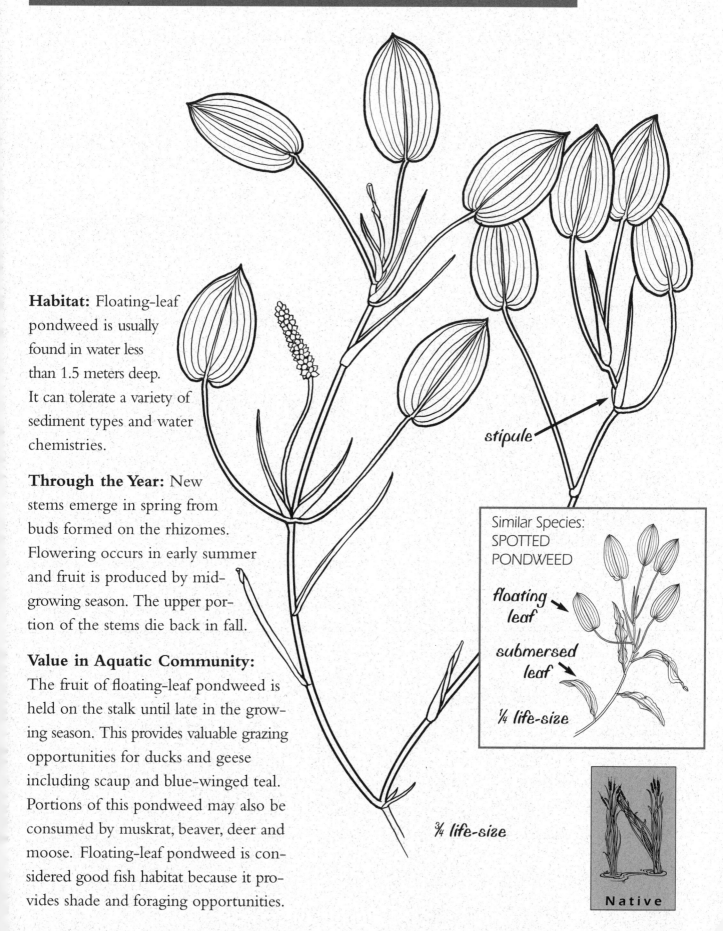

Habitat: Floating-leaf pondweed is usually found in water less than 1.5 meters deep. It can tolerate a variety of sediment types and water chemistries.

Through the Year: New stems emerge in spring from buds formed on the rhizomes. Flowering occurs in early summer and fruit is produced by mid-growing season. The upper portion of the stems die back in fall.

Value in Aquatic Community: The fruit of floating-leaf pondweed is held on the stalk until late in the growing season. This provides valuable grazing opportunities for ducks and geese including scaup and blue-winged teal. Portions of this pondweed may also be consumed by muskrat, beaver, deer and moose. Floating-leaf pondweed is considered good fish habitat because it provides shade and foraging opportunities.

stipule

Similar Species:
SPOTTED PONDWEED

floating leaf

submersed leaf

¼ *life-size*

¾ *life-size*

Native

Potamogeton nodosus (POT-a-mo-JEE-ton no-DOE-sus)

Long-leaf pondweed

Potamogeton – (Gk.) *potamos:* river + *geiton:* neighbor; *nodosus* – (L.) knotty

*The flowing water buoys the long leaves of the pondweed.
The plant reaches an amazing length. A closer look reveals
a miniature metropolis of curious creatures straight out of a
Star Wars movie . . . tiny teeth, tentacles, creepers and crawlers
living out their lives on this leafy abode.*

Description: Long-leaf pondweed has stems (up to 2 m long) that emerge from branching rhizomes. Submersed leaves (up to 30 cm long, 1-2.5 cm wide) are narrowly lance-shaped and gradually taper to a long leaf stalk. The floating leaves (5-13 cm long, 1-4 cm wide) also taper to long leaf stalks. Stipules (4–10 cm long) of both the submersed and floating leaves are free in the axils of the leaves and gradually break down over the growing season. (See *Potamogeton* spp. for a description of stipules.)

Flowers and fruit are produced on a dense, cylindrical spike (3–5 cm long) that pokes up above the water surface. The fruits (3-4 mm long) are somewhat oval in outline with a lumpy dorsal ridge, a short beak and sometimes low lateral ridges.

Similar species: The long leaf stalks of the submersed leaves help separate long-leaf pondweed from species with similar floating leaves such as Illinois pondweed (*P. illinoensis*) and floating-leaf pondweed (*P. natans*).

Origin & Range: Native; found at scattered locations throughout Wisconsin; range includes most of U.S.

Habitat: Long-leaf pondweed is more common in flowing water than lakes. It is often found in water about 1 meter deep on a variety of sediment types. It can tolerate turbid water and is often associated with other species that do well in eutrophic situations.

Through the Year: Long-leaf pondweed overwinters by hardy rhizomes. New stems emerge in spring and flowering occurs by early summer. Fruits are produced in mid- to late summer. Foliage usually dies back in the fall, but under some conditions parts of the plant may overwinter as green shoots.

Value in the Aquatic Community: The fruit of long-leaf pondweed is grazed by a variety of ducks and geese. Portions of the plant are eaten by muskrat, beaver, deer and moose. Long-leaf pondweed offers invertebrate habitat and foraging opportunities for fish.

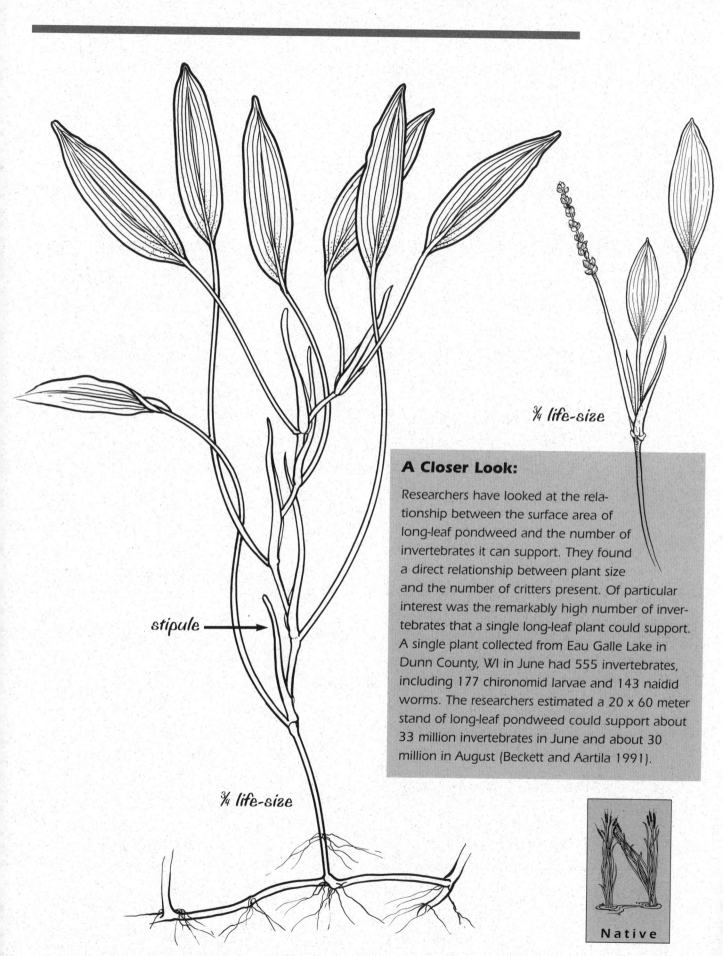

stipule

¾ life-size

¾ life-size

A Closer Look:

Researchers have looked at the relationship between the surface area of long-leaf pondweed and the number of invertebrates it can support. They found a direct relationship between plant size and the number of critters present. Of particular interest was the remarkably high number of invertebrates that a single long-leaf plant could support. A single plant collected from Eau Galle Lake in Dunn County, WI in June had 555 invertebrates, including 177 chironomid larvae and 143 naidid worms. The researchers estimated a 20 x 60 meter stand of long-leaf pondweed could support about 33 million invertebrates in June and about 30 million in August (Beckett and Aartila 1991).

Native

Potamogeton pectinatus (POT-a-mo-JEE-ton PECK-tin-a-tus)

Sago pondweed

Potamogeton – (Gk.) *potamos*: river + *geiton*: neighbor;
pectinatus – (L.) comb-like (referring to narrow leaves)

With a snarl the outboard motor kicked out of the water. The lake was so murky that the skiers didn't realize how shallow it had become. The hushed prop was tangled in a green mass that had been sago pondweed.

Similar Species:
SHEATHED
PONDWEED

inflated
stipular
sheath

⅛ life-size

Description: The stems (30–80 cm long) of sago pondweed sprout from slender rhizomes that are peppered with starchy tubers (1–1.5 cm long). The leaves (3–10 cm long, 0.5–1.5 mm wide) are very thin and resemble pine needles, ending in a sharp point. Each branch may be forked several times into a spreading, fan–like arrangement. Stipules are fused to the leaves for most of their length, creating a stipular sheath (1–3 cm long). (See *Potamogeton* spp. for a description of stipules.)

stipular
sheath
6 x life-size

Flowers and fruit are produced on a slender stalk (3–10 cm long) that may be submersed or floating on the water surface. The flowers and fruit are arranged in small whorls that are slightly spaced apart on the stalk. This creates the appearance of beads on a string. Each fruit (3–4.5 mm long) is oval to egg-shaped in outline. The fruit is rounded on the back with a short beak and sometimes a low dorsal ridge, two lateral ridges or both.

Similar species: Sago pondweed resembles two other pondweeds with needle-like leaves, but is much more common than either of them.

Thread-leaf pondweed (*P. filiformis*) has leaves with a blunt, notched tip and a smaller fruit than sago pondweed.

Sheathed pondweed (*P. vaginatus*) also has a blunt, notched tip and is easily distinguished from sago by the inflated sheaths of its stipules (2.5–5 mm wide). Sheathed pondweed is listed as **Threatened** in Wisconsin.

Origin & Range: Native; common in Wisconsin; range includes most of U.S.

Habitat: Sago pondweed is widespread in lakes, ponds, streams and rivers. It is usually found in water 1–2 meters deep, but will grow much deeper at times.

SAGO PONDWEED

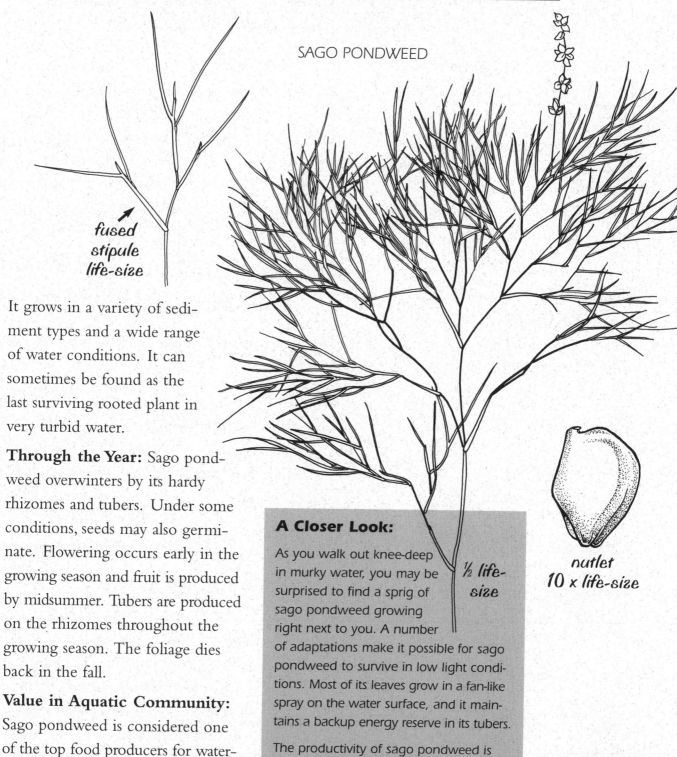

fused
stipule
life-size

½ life-size

nutlet
10 x life-size

It grows in a variety of sediment types and a wide range of water conditions. It can sometimes be found as the last surviving rooted plant in very turbid water.

Through the Year: Sago pondweed overwinters by its hardy rhizomes and tubers. Under some conditions, seeds may also germinate. Flowering occurs early in the growing season and fruit is produced by midsummer. Tubers are produced on the rhizomes throughout the growing season. The foliage dies back in the fall.

Value in Aquatic Community: Sago pondweed is considered one of the top food producers for waterfowl. Both the fruit and tubers are heavily grazed and are considered critical for a variety of migratory waterfowl. Sago also provides food and shelter for young trout and other juvenile fish.

A Closer Look:

As you walk out knee-deep in murky water, you may be surprised to find a sprig of sago pondweed growing right next to you. A number of adaptations make it possible for sago pondweed to survive in low light conditions. Most of its leaves grow in a fan-like spray on the water surface, and it maintains a backup energy reserve in its tubers.

The productivity of sago pondweed is remarkable. A single plant grown in cultivation produced 63,000 fruits and 36,000 tubers over a 6-month period (Voss 1972). Because it is so valuable for waterfowl, sago has been planted in shallow lakes and wildlife ponds to create additional duck habitat.

Native

SUBMERSED

Potamogeton praelongus (POT-a-mo-JEE-ton pray-LON-gus)

White-stem pondweed

Potamogeton – (Gk.) *potamos*: river + *geiton*: neighbor;
praelongus – (L.) greatly prolonged

The water was gin clear. The pale stems of white-stem pondweed zigzagged toward the surface like bent soda straws. The boat-shaped tips of the clasping leaves looked like miniature canoes.

Description: The zigzag stems (2–3 m long, 1.5–4 mm thick) of white-stem pondweed emerge from a stout, rust-spotted rhizome. Submersed leaves (8–30 cm long, 1–4.5 cm wide) are lance- to oval-shaped and clasp the stem, wrapping around one-third to one-half the stem's diameter. The leaves have 3–5 strong veins and many (11–35) weaker ones. The tip of the leaf is boat-shaped and splits when pressed, creating a notch at the end of the leaf.

boat-shaped
leaf tip
life-size

Stipules (3–8 cm) are white and fibrous, often shredding at the tip over the growing season. They are free in the leaf axils, but are usually closely pressed against the stem. (See *Potamogeton* spp. for definition of stipules.) No floating leaves are produced.

Flowers and fruits are arranged in a cylindrical spike (3–7.5 cm long) that may be continuous or interrupted. Fruit (4–5.7 mm) is oval to egg-shaped in outline and rather plump. The surface of the fruit has a short beak, a sharp dorsal ridge and often two low lateral ridges.

Similar species: White-stem pondweed could be confused with clasping-leaf pondweed (*P. richardsonii*). However, clasping-leaf has smaller leaves that are not boat-shaped at the tip, and the fruit does not have a sharp dorsal ridge.

Origin & Range: Native; common in northern and eastern Wisconsin; range includes northern and western U.S.

Habitat: White-stem pondweed is usually found in soft sediment in water ranging from 1–4 meters deep. It is generally found in lakes with good water clarity.

Through the Year: White-stem pondweed generally overwinters by hardy rhizomes but under some conditions it may remain evergreen under the ice. Flowering and fruiting occur by mid-summer. The fruit is held on the stalk until late in the summer.

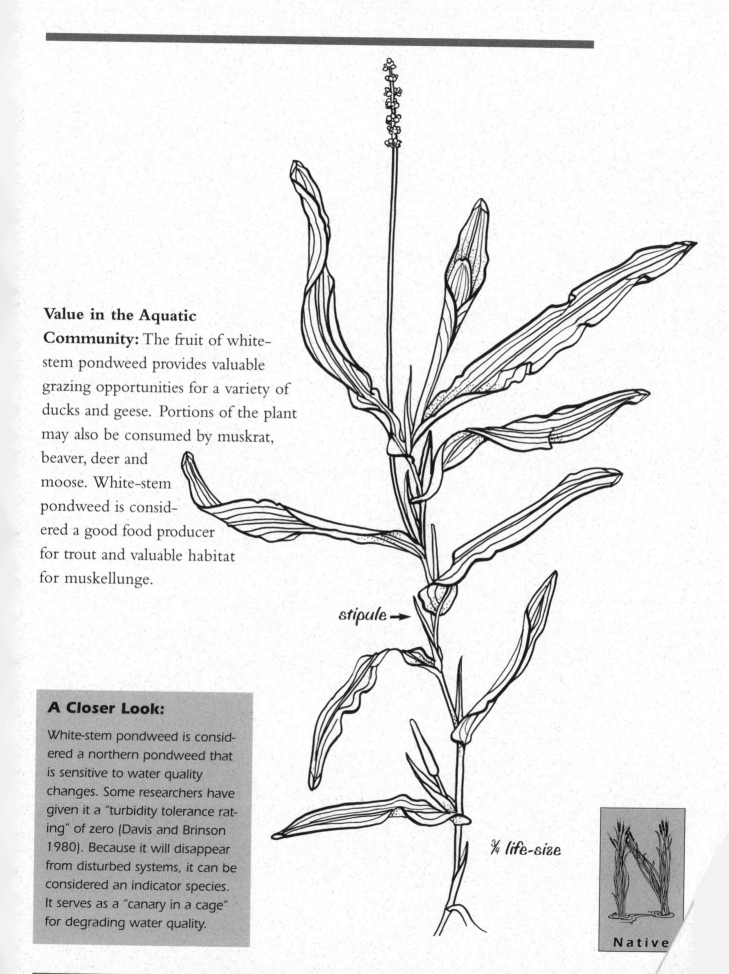

Value in the Aquatic Community: The fruit of white-stem pondweed provides valuable grazing opportunities for a variety of ducks and geese. Portions of the plant may also be consumed by muskrat, beaver, deer and moose. White-stem pondweed is considered a good food producer for trout and valuable habitat for muskellunge.

stipule →

A Closer Look:

White-stem pondweed is considered a northern pondweed that is sensitive to water quality changes. Some researchers have given it a "turbidity tolerance rating" of zero (Davis and Brinson 1980). Because it will disappear from disturbed systems, it can be considered an indicator species. It serves as a "canary in a cage" for degrading water quality.

¾ life-size

Native

Potamogeton pusillus (POT-a-mo-JEE-ton pu-SIL-us)

Small pondweed

Potamogeton – (Gk.) *potamos:* river + *geiton:* neighbor; *pusillus* – (L.) very small

The stillness is shattered as a squadron of mallards take to the air. Each morning about this time, the moose cow and her calf wade into the shallow bay to feed on the small pondweeds growing there.

Description: The slender stems (up to 1.5 m long) of small pondweed emerge from a slight rhizome and branch repeatedly near the ends. Submersed leaves (1-7 cm long, 0.2-2.5 mm wide) are linear and attach directly to the stem. There are usually a pair of raised glands at the base of the leaf attachment. The leaves have three veins and the mid-vein may be bordered by several rows of lacunar (hollow) cells. The tip of the leaf varies from blunt to pointed.

Membranous stipules (5-15 mm long) may be wrapped around the stem early in the season, but gradually break down. (See *Potamogeton* spp. for description of stipules.) Numerous winter buds are usually produced. These buds (1-3 cm long) have the inner leaves rolled into a tight cigar

mid-vein bordered by lacunar cells 6 x life-size

shape. No floating leaves are produced.

The flowers and fruits are produced in 1-4 whorls on a slender stalk (0.5-6 cm long). The oval fruit (1.5-2.2 mm long) is rather plump and has a smooth back and short beak.

nutlet 10 x life-size

Similar species: The fine-leaved pondweeds of the "pusilli" group are a bit like sparrows – you know there are many species, but they are hard to separate. With sparrows the songs are unique; with the small pondweeds the fruit and winter buds are distinctive. The most common fine-leaved pondweed of this group is its namesake, *Potamogeton pusillus*. It grows in a wide range of habitats and is a key element of many aquatic plant communities. One other member of this group, *Potamogeton berchtoldii*, is now combined with *P. pusillus* by a number of taxonomists.

(continued)

SMALL
PONDWEED

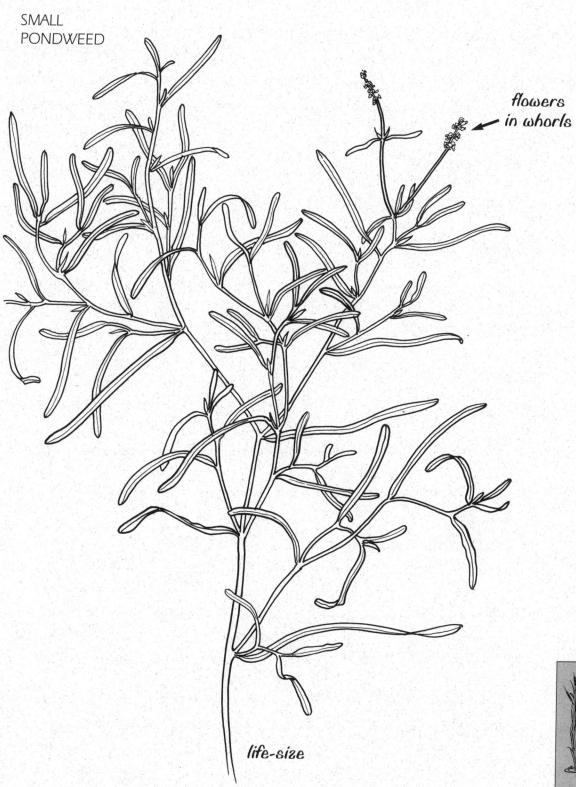

flowers
in whorls

life-size

Native

Potamogeton pusillus (continued)

Small pondweed could be confused with leafy pondweed. Leafy pondweed (*P. foliosus*) has leaves that are similar to small pondweed, but there are no glands at the leaf nodes. The short fruiting stalk of leafy pondweed has a tight cluster of fruits with sharp dorsal ridges (see *P. foliosus* description). Winter buds are 1-2 cm long with tightly-rolled inner leaves.

The following is a brief description of other members of this group and how they can be distinguished from small pondweed:

Hill's pondweed (*P. hillii*) has narrow leaves (3-7 cm long, 1-2 mm wide) that are usually bristle-tipped. In most cases, there are no glands present. The stipules are white and fibrous rather than membranous. Recurved fruiting stalks (5-15 mm long) develop in the upper leaf axils. The fruit has both a low, sharp dorsal ridge and lateral ridges. Winter buds are hard and fibrous at the base.

Blunt-leaf pondweed (*P. obtusifolius*) has leaves (3-10 cm long, 1-3.5 mm wide) that are usually longer and wider than small pondweed. The leaves are often tinged with red and blunt-tipped. Blunt-leaf pondweed

has nodal glands and membranous stipules that are similar to small pondweed, but the fruit is produced on a short stalk and has a low, sharp dorsal ridge. Winter buds (3.5-8 cm long) are frequently produced, but the leaves are not strongly modified.

Fries' pondweed (*P. friesii*) has narrow leaves (3-7 cm long, 1.5-3 mm wide) that each have 5-7 veins and a rounded tip with a short beak. There are a pair of glands at the nodes and white, fibrous stipules. Flowers and fruits are borne on a flattened stalk (1.5-5 cm long). The fruit (2-3 mm long) is rounded with low ridges or no ridges. Winter buds (1.5-5 cm

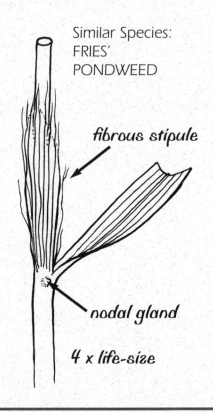

Similar Species:
FRIES'
PONDWEED

fibrous stipule

nodal gland

4 x life-size

long) are a strong characteristic. The inner leaves of each winter bud are compressed and arranged in a fan shape that is at a right angle to the outer leaves.

Stiff pondweed (*P. strictifolius*) has narrow, stiff leaves (2-6 cm long, 0.5-2 mm wide) that have 3-5 veins and a pointed tip. The stipules are white and fibrous. In most cases, there are no glands present. The slender flower stalk (3-4 cm long) produces whorls of rounded fruit (each 2-3 mm long). Winter buds are flattened with the outer leaves arching outward.

Origin & Range: Native; common throughout Wisconsin; range includes most of U.S.

Habitat: Small pondweed will tolerate turbid conditions and is found from shallow zones to water 2-3 meters deep.

Through the Year: Small pondweed overwinters by rhizomes and winter buds. There is some limited reproduction by seed when conditions are right. Flowering occurs early in the growing season and fruit is produced by mid-season. Stems and leaves die back in the fall, leaving fragments with winter buds on the sediment.

Value in Aquatic Community: Small pondweed can be a locally important food source for a variety of ducks and geese including gadwall, mallard, northern pintail, ring-necked duck, white-winged scoter, blue-winged teal, green-winged teal and American wigeon. The plant may also be grazed by muskrat, deer, beaver and moose. Small pondweed provides a food source and cover for fish.

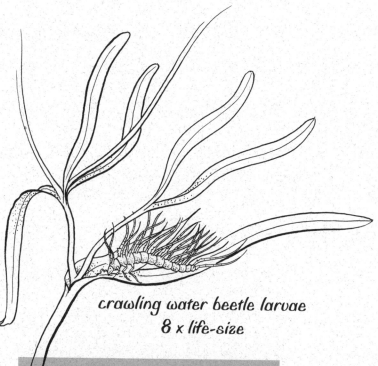

crawling water beetle larvae
8 x life-size

A Closer Look:

If you take a look at a piece of small pondweed through a magnifying glass, you may discover the larva of a crawling water beetle. Both the adults and larvae are common in the vegetation of shallow water, and can be seen grazing on the leaves or stipules of small pondweed.

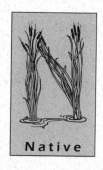

Native

Potamogeton richardsonii
(POT-a-mo-JEE-ton RICH-ard-SON-ee-i)

Clasping-leaf pondweed

Potamogeton – (Gk.) *potamos:* river + *geiton:* neighbor;
richardsonii – named for Sir John Richardson, a Scottish naturalist (1787-1865)

*It had been years since anyone had visited the little pond.
The city had grown up around it. Abandoned cars and rubble
lined its banks and the water was too murky to support
many plants. The clasping-leaf pondweed was able to succeed
there – a detail of which many ducks were well aware.*

Description: Clasping-leaf pondweed has sinuous stems (1-2.5 mm thick) that emerge from a spreading rhizome. Oval to somewhat lance-shaped leaves (3-12 cm long, 0.5-2 cm wide) clasp the stem. The base of each leaf is heart-shaped and covers one-half to three-quarters of the stem circumference. Leaves have 13-21 veins (some more prominent than others). The axil of each leaf has a fibrous stipule that soon disintegrates, leaving a beard of white fibers at the leaf node. No floating leaves are produced.

Fruiting stalks (1.5-25 cm long) develop in the upper leaf axils. The cylindrical spikes (1.5-3 cm long) are packed with fruit. Each olive-green fruit (2.2-4.2 mm long) is plump and round with a prominent beak (1 mm long).

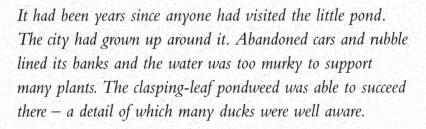

**nutlet
10 x life-size**

Similar species: Clasping-leaf pondweed most closely resembles **redhead pondweed** (*Potamogeton perfoliatus*) or white-stem pondweed (*Potamogeton praelongus*). Redhead pondweed can be distinguished by its smaller leaves that completely wrap the stem at their base. White-stem pondweed can be separated by the boat-shaped tips of the leaves. The foliage of curly-leaf pondweed (*Potamogeton crispus*) may also resemble clasping-leaf, but the margins of curly-leaf are serrated.

Origin & Range: Native; common throughout Wisconsin; range includes most of U.S.

Habitat: Clasping-leaf pondweed can be found growing in a variety of sediment types in water up to 4 meters deep. It is tolerant of disturbance and is often found growing with coontail (*Ceratophyllum demersum*) and small pondweed (*Potamogeton pusillus*).

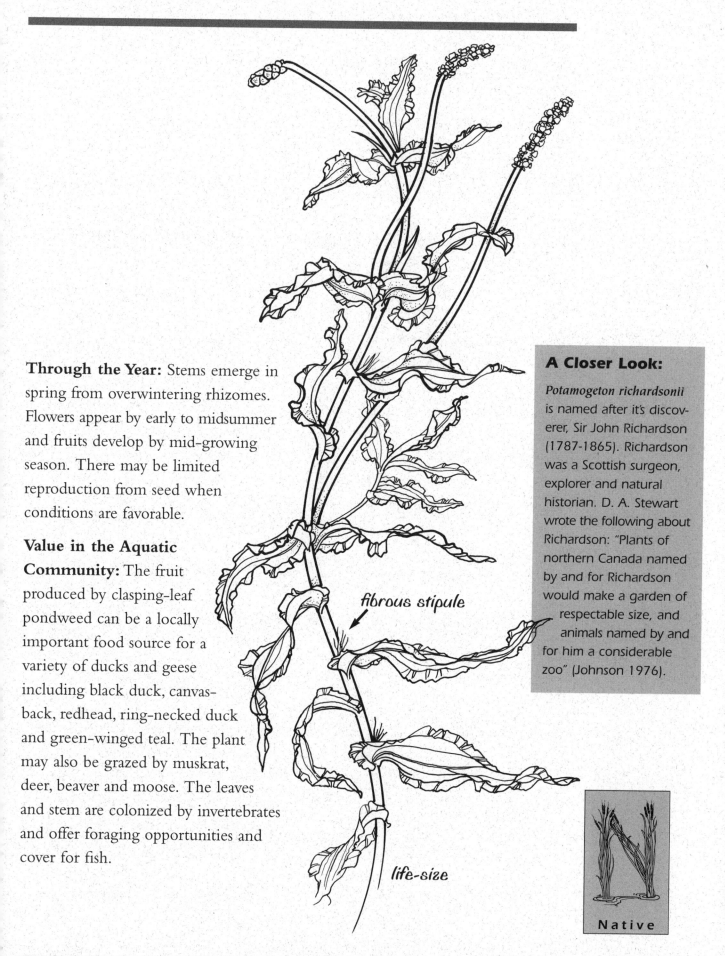

Through the Year: Stems emerge in spring from overwintering rhizomes. Flowers appear by early to midsummer and fruits develop by mid-growing season. There may be limited reproduction from seed when conditions are favorable.

Value in the Aquatic Community: The fruit produced by clasping-leaf pondweed can be a locally important food source for a variety of ducks and geese including black duck, canvasback, redhead, ring-necked duck and green-winged teal. The plant may also be grazed by muskrat, deer, beaver and moose. The leaves and stem are colonized by invertebrates and offer foraging opportunities and cover for fish.

fibrous stipule

life-size

A Closer Look:

Potamogeton richardsonii is named after it's discoverer, Sir John Richardson (1787-1865). Richardson was a Scottish surgeon, explorer and natural historian. D. A. Stewart wrote the following about Richardson: "Plants of northern Canada named by and for Richardson would make a garden of respectable size, and animals named by and for him a considerable zoo" (Johnson 1976).

Native

SUBMERSED

Potamogeton robbinsii (POT-a-mo-JEE-ton row-BINS-ee-i)

Fern pondweed, Robbins pondweed

Potamogeton – (Gk.) *potamos*: river + *geiton*: neighbor;
robbinsii – named for James W. Robbins (1801–1879)

Swimming out from shore, the warm sand bottom gives way to the shadowy green sweep of a drop-off. Along a deep incline, fern pond-weed rises in prominent beds forming a luxuriant meadow of foliage. This is the domain of the northern pike that patrol the edge.

Description: Robust stems of fern pondweed emerge from a spreading rhizome. The leaves are strongly two-ranked, creating a feather or fern-like appearance which is most evident when the plant is still in the water. Each leaf (3-10 cm long, 3-8 mm wide) is firm and linear, with a base that wraps around the stem. The leaf base is distinctive. It has small ear-like lobes at the juncture with the stipule and is fused with the fibrous stipule for about 5-15 mm. This creates a tufted appearance just up from the leaf base. The leaves are closely spaced and have a finely serrated margin. No floating leaves are produced.

Flowering stalks (2-5 cm long) develop in the upper leaf axils. Whorls of flowers are produced, but fruit rarely develops. When fruit (3.5-4.5 mm) does mature, it has a sharp dorsal ridge and two rounded lateral ones. Winter buds are often produced on terminal branches. These buds have tightly compressed, reduced leaves.

serrated leaf margin 2 x life-size

Similar species: The strongly two-ranked leaves and fern-like appearance give fern pondweed a unique look that is unlikely to be confused with other species.

Origin & Range: Native; found primarily in northern and eastern Wisconsin; range includes most of U.S.

Habitat: Fern pondweed thrives in deeper water than any other pondweed in our region. It is not unusual to find a ring of fern pondweed that follows a depth contour just beyond the outer margin of a mixed stand of plants.

Through the Year: Fern pondweed sprouts in the spring from rhizomes and winter buds. Flowering occurs by midsummer but fruit is rarely produced. Stems and leaves usually die back in the fall, leaving fragments with winter buds on the sediment. Under some conditions, portions of fern pondweed may overwinter green.

Value in the Aquatic Community: Fern pondweed provides habitat for invertebrates that are grazed by waterfowl. It also offers good cover and foraging opportunities for fish, particularly northern pike.

nutlet
10 x life-size

¾ life-size

A Closer Look:

The seasonal growth pattern of *Potamogeton robbinsii* was studied in Lake George, New York. Annual density ranged from 100-1,000 plants per square meter in the 5-7 meter depth zone. This density level remained quite constant throughout the year. Plants gathered during the winter showed 10-20% the photosynthetic activity of plants gathered during midsummer (Boylen and Sheldon 1976).

Native

Potamogeton spirillus (POT-a-mo-JEE-ton spi-RIL-us)

Spiral-fruited pondweed

Potamogeton – (Gk.) *potamos*: river + *geiton*: neighbor; *spirillus* – (L.) *spira*: a coil

The sandy cove was a busy place. The ducks were hard at work. Heads down, bottoms up, they deftly nibbled the nutlike fruits that grew in clusters on the spiral-fruited pondweed.

SUBMERSED

Description: Spiral-fruited pondweed has slender stems that emerge from a slight, spreading rhizome. The stems have a much-branched, compact form. Submersed leaves are linear (1-8 cm long, 0.5-2 mm wide) and usually have a curved appearance, as though they had been curled around a pencil. Stipules are fused to the leaf blade for more than half their length. (See *Potamogeton* spp. for definition of stipules.) When present, the floating leaves are shaped like an ellipse (7-35 mm long, 2-13 mm wide).

nutlet with spiral-coiled embryo

10 x life-size

The nutlike fruits (1.3-2.4 mm) are produced on stalks of varying lengths. Those in the lower leaf axils have short stalks and fruit in a compact head. Those in upper leaf axils have longer stalks and a more cylindrical cluster of fruit. Each fruit looks like a flattened disc. There is a sharply toothed margin on the top edge of the fruit and smooth sides that tightly conform to the spiral-coiled embryo.

Similar species: Spiral-fruited pondweed can be confused with other narrow-leaved pondweeds. Fortunately, the distinctive fruit is produced early in the growing season and is the best distinguishing feature.

Origin & Range: Native; found primarily in northern Wisconsin; range includes northern U.S.

Habitat: Spiral-fruited pondweed is usually found in shallow water. It will grow in a variety of sediment types.

Through the Year: Spiral-fruited pondweed overwinters by rhizomes and winter buds. There is some limited reproduction by seed when conditions are right. Flowering occurs early in the growing season and fruit is produced by mid-season. Stems and leaves die back in the fall, leaving fragments with winter buds on the sediment.

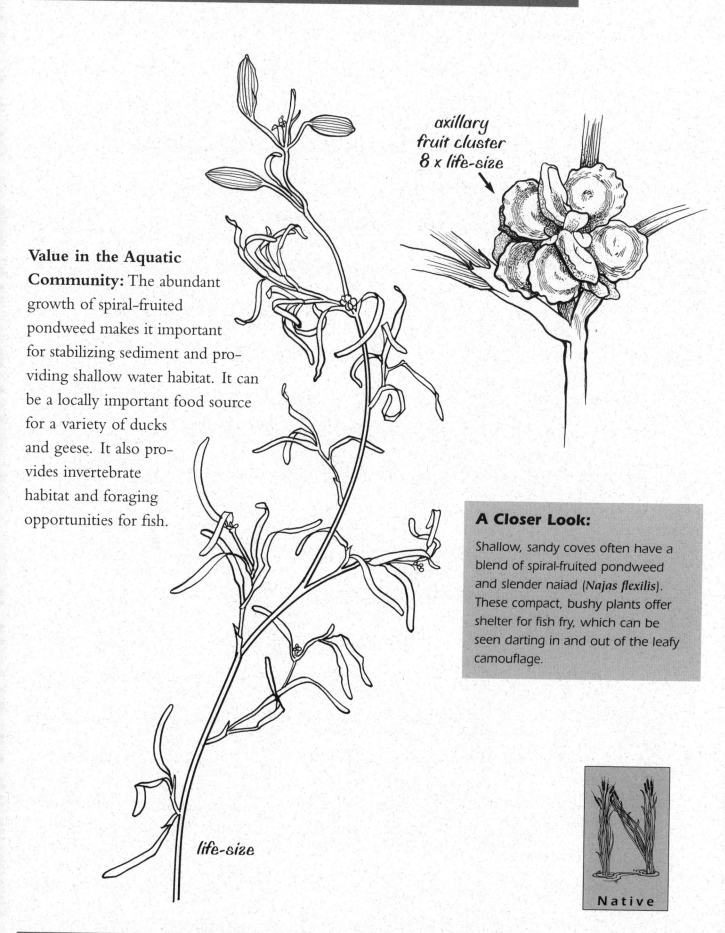

axillary
fruit cluster
8 x life-size

Value in the Aquatic Community: The abundant growth of spiral-fruited pondweed makes it important for stabilizing sediment and providing shallow water habitat. It can be a locally important food source for a variety of ducks and geese. It also provides invertebrate habitat and foraging opportunities for fish.

A Closer Look:

Shallow, sandy coves often have a blend of spiral-fruited pondweed and slender naiad (*Najas flexilis*). These compact, bushy plants offer shelter for fish fry, which can be seen darting in and out of the leafy camouflage.

life-size

Native

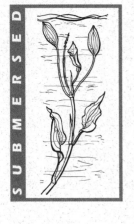

SUBMERSED

Potamogeton zosteriformis
(POT-a-mo-JEE-ton ZOS-ter-i-FORM-is)

Flat-stem pondweed

Potamogeton – (Gk.) *potamos:* river + *geiton:* neighbor; *zosteriformis* – (L.) *zostera:* eelgrass (a marine plant with flat leaves) + *formis:* having the form of

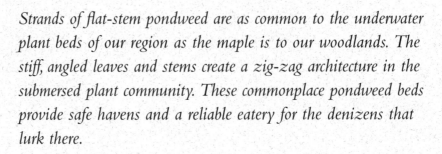

Strands of flat-stem pondweed are as common to the underwater plant beds of our region as the maple is to our woodlands. The stiff, angled leaves and stems create a zig-zag architecture in the submersed plant community. These commonplace pondweed beds provide safe havens and a reliable eatery for the denizens that lurk there.

Description: Freely-branched stems of flat-stem pondweed emerge from a slight rhizome. The stems are strongly flattened and have an angled appearance. Stiff linear leaves (10-20 cm long, 2-5 mm wide) have a prominent midvein and many fine, parallel veins. The firm stipules (1-2 cm long) are free in the leaf axils. (See *Potamogeton* spp. for a description of stipules.) No floating leaves are produced.

The nutlike fruits are arranged in a cylindrical spike that pokes out of the water. Each oval fruit (4-4.5 mm) has a sharp, narrow dorsal ridge. Winter buds (4-7.5 cm) are commonly formed. They have a cluster of tightly packed ascending leaves, with the leaf tips extending beyond the fibrous stipules.

nutlet
8 x life-size

Similar species: Flat-stem pondweed is most commonly confused with water stargrass (*Zosterella dubia*). The size and arrangement of the leaves is similar, but the leaves of water stargrass lack a prominent midvein. Water stargrass also has very different flowers and fruit. The flowers are yellow with narrow petals and the fruits are capsules.

Origin & Range: Native; common throughout Wisconsin; range includes northern and western U.S.

Habitat: Flat-stem pondweed grows in a variety of water depths from shallow to several meters deep. It is usually found in soft sediment.

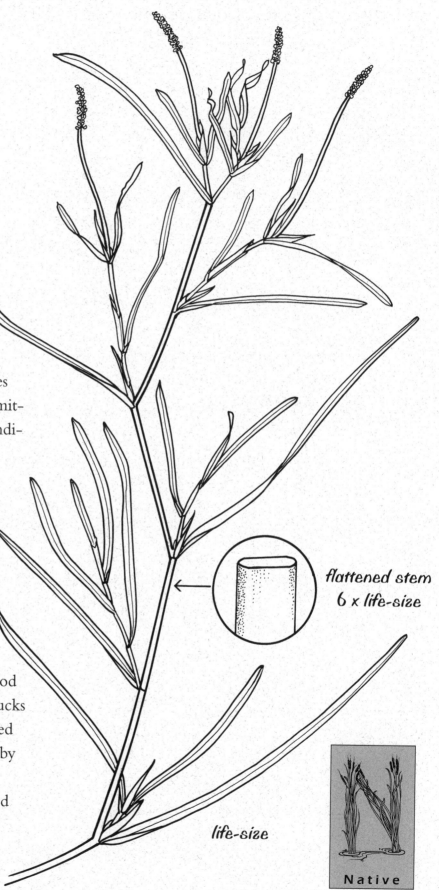

A Closer Look:

Laboratory tests have shown that flat-stem pondweed has high antimicrobial properties. This resistance may contribute to its vigor in many plant populations (Gopal and Goel 1993).

Through the Year: Flat-stem pondweed overwinters by rhizomes and winter buds. There is some limited reproduction by seed when conditions are right. Flowering occurs early in the growing season and fruit is produced by mid-season. Stems and leaves die back in the fall, leaving fragments with winter buds on the sediment.

Value in the Aquatic Community: Flat-stem pondweed can be a locally important food source for a variety of geese and ducks including redhead and green-winged teal. The plant may also be grazed by muskrat, deer, beaver and moose. Flat-stem pondweed provides a food source and cover for fish and invertebrates.

*flattened stem
6 x life-size*

life-size

Native

Ranunculus flammula (rah-NUN-cue-les FLAM-u-la)

(formerly known as Ranunculus reptans L.)

Creeping spearwort

Ranunculus – (L.) little frog; *flammula* – (L.) small flame
(referring to the burning, acrid juice)

The small tufts of spearwort look like they are playing "leap frog." They are connected by arched runners that hopscotch across the sediment and take root at each node, producing a few grass-like leaves.

Description: Linear leaves (up to 7 cm long, 1-4 mm wide) emerge in small clusters from the creeping surface stem, called a runner or stolon. The runners are about the same width as the leaves. The leaves are not tapered and have a blunt tip.

In shallow water or on the shoreline, spearwort may produce flowers. A single-flowered stem (3-15 cm) emerges at some nodes. Each flower has five yellow petals (2-4 mm long) that are about twice as long as the sepals. A cluster of nutlets (1.3-1.7 mm long) with short, erect beaks (0.2-0.5 mm) are produced as the flower matures.

Similar Species: The submersed form of creeping spearwort could be confused with other turf-forming submersed plants such as brown-fruited rush (*Juncus pelocarpus* f. *submersus*) or needle spike-rush (*Eleocharis acicularis*). Brown-fruited rush has rhizomes that are much thinner than the leaves and the leaves are cupped around one another. Needle spikerush also has very fine rhizomes and the tufts are stems that end in a pointed tip or small spikelet.

Origin & Range: Native; scattered locations in northern and central Wisconsin; range includes northern U.S.

Habitat: Creeping spearwort is found primarily on sand or gravel substrate in soft-water lakes. It grows from moist shorelines to water 2 meters deep.

Through the Year: Leaves resprout from the rootstalks in early spring. Flowering occurs by early to midsummer and nutlets mature by late summer.

Value in the Aquatic Community: Sprawling submersed beds of spearwort offer valuable invertebrate habitat and fish spawning areas.

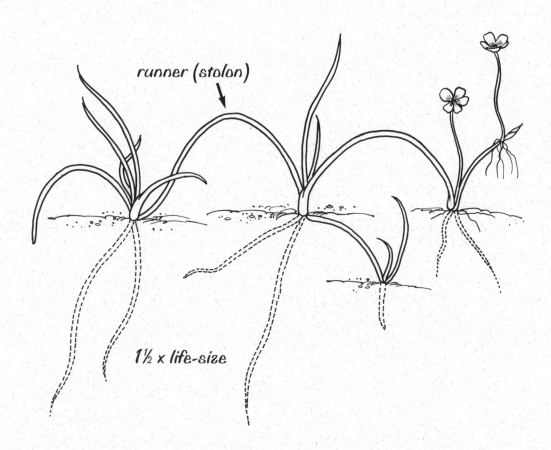

runner (stolon)

1½ × life-size

A Closer Look:

The Latin meaning of *Ranunculus* is "little frog." The "leap frog" growth form of creeping spearwort makes the name particularly appropriate.

Native

SUBMERSED

Scirpus subterminalis (SKER-pus sub-ter-min-AL-es)

Water bulrush

Scirpus – (L.) bulrush; *subterminalis* – (L.) *sub:* below + *terminalis:* end, terminal

*The basking turtles were startled when the blue heron
descended on the huge pine. They plopped off and paddled
frantically through the limp stems of the water bulrush.
Their headlong charge ended in a cloud as they dove into
the sanctuary of the soft sediments.*

Description: Water bulrush is the most truly aquatic bulrush in our region with only the tips of fertile stems poking out of the water. Stems develop from a fine rhizome (less than 2 mm diameter). Slender, limp stems (up to 1 m long) float in the water along with hair-like leaves that arise near the base.

The leaves sheath one another at the base and are crescent-shaped just above the sheath. The leaves have 1-5 lengthwise veins with scattered cross-veins. A solitary spikelet (7-12 mm long) may be produced on a fertile stem. A floral leaf (1.5-6 cm long) extends above the spikelet like an extension of the stem. The scales of the spikelet are light brown and very thin. Nutlets develop beneath the scales. Each nutlet (2.5-3.8 mm) is three-angled with a slender beak.

Similar Species: Submersed leaves of water bulrush could be confused with the fine, submersed stems of Robbins spikerush (*Eleocharis robbinsii*). However, the leaf-like stems of Robbins spikerush are all separate, while the leaves of water bulrush sheath each other at the base.

Origin & Range: Native; scattered locations in Wisconsin; range includes most of U.S.

Habitat: Water bulrush can be found in both still and flowing water in a variety of sediment types. It is often growing in water up to 1 meter or more deep.

Value in the Aquatic Community: Grass-like meadows of water bulrush provide invertebrate habitat and shelter for fish.

A Closer Look:

Researchers have studied the fine coating of algae that grows on water bulrush stems and leaves in phosphorus-limited lakes. They found the bulrush was a source of phosphorus for the algae throughout the growing season. In nutrient-poor lakes, limited algae growth can in turn limit invertebrate and fish growth. The phosphorus transfer from water bulrush to the microflora is a valuable link in this system (Burkholder and Wetzel 1990).

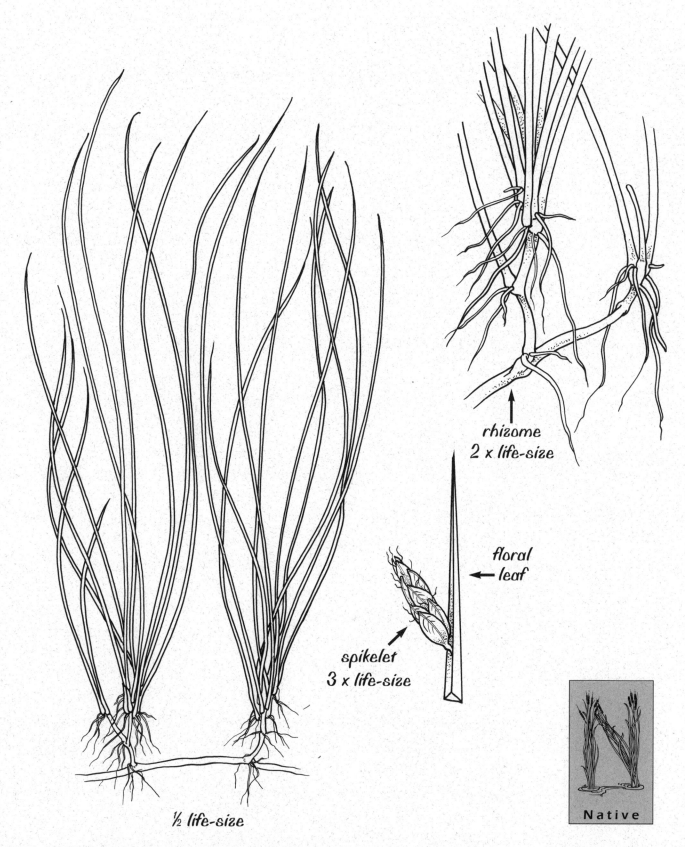

rhizome
2 x life-size

floral
leaf

spikelet
3 x life-size

½ life-size

Native

SUBMERSED

Vallisneria americana (VAL-is-NER-ee-a a-mer-e-KAN-a)

Wild celery, eel-grass, tape-grass

Vallisneria – named for Antonio Vallisneri, an Italian botanist (1661-1730);
americana – American

The steel blue clouds and crisp morning air foreshadowed the cold to come. Great flocks of canvasbacks circled and landed in the beds of wild celery. They dove among the trailing cellophane leaves in search of the thick, starchy tubers that would boost their energy for the long journey ahead.

Description: Wild celery has ribbon-like leaves that emerge in clusters along a creeping rhizome. The leaves (up to 2 m long, 3-10 mm wide) have a prominent central stripe and a cellophane-like consistency. The leaves are mostly submersed, with just the tips trailing on the surface of the water.

Male and female flowers are produced on separate plants. The tiny male flowers (1 mm wide) are clustered in a case that develops underwater. As the flowers mature, they are released from the case. Each male flower is in a closed "floral envelope" that contains an air bubble. This helps lift it to the surface. When it reaches the surface, the floral envelope opens and creates a sail that allows it to skim along the surface.

The female flowers (3.5-6.5 mm wide) also develop underwater, but then are raised to the surface by a fast-growing, spiral-coiled stalk. These delicate, white flowers bob at the surface creating a dip in the surface tension. When one of the

tiny male flowers sails by, it glides down to meet and pollinate the female flower. After fertilization, the female flower is retracted beneath the surface and a long, capsular fruit (5-12 cm) develops.

Similar Species: The foliage of wild celery is sometimes confused with the submersed leaves of bur-reeds (*Sparganium* spp.) or arrowheads (*Sagittaria* spp.). The prominent middle stripe will usually distinguish wild celery. The leaves of ribbon-leaf pond-weed (*Potamogeton epihydrus*) are also similar, but they are alternate on a stem rather than basal.

Origin & Range: Native; found throughout Wisconsin; range includes most of U.S.

Habitat: Wild celery is usually found growing in firm substrates in water ranging from ankle-deep to several meters. It is turbidity tolerant and will survive in a broad range of water chemistries.

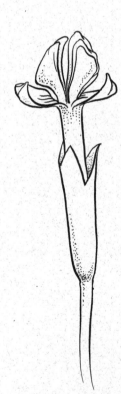

*female flower
4 x life-size*

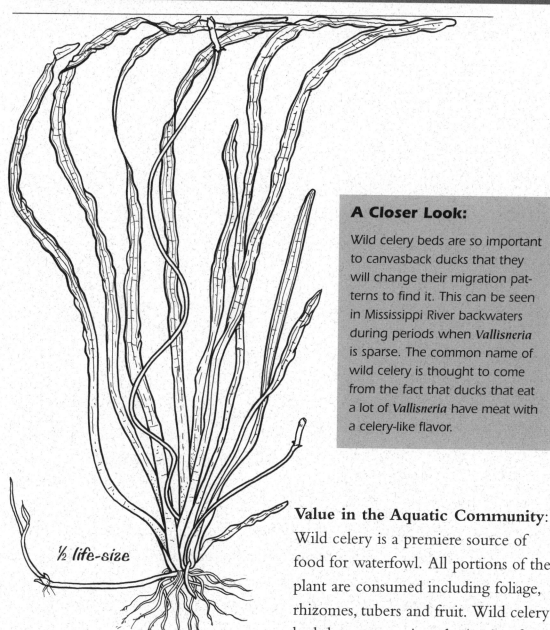

½ life-size

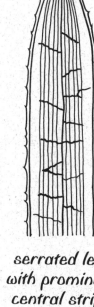

serrated leaf
with prominent
central stripe
2 x life-size

Value in the Aquatic Community: Wild celery is a premiere source of food for waterfowl. All portions of the plant are consumed including foliage, rhizomes, tubers and fruit. Wild celery beds become a prime destination for thousands of canvasback ducks every fall. The relationship between wild celery and canvasbacks is so strong that the scientific name for these ducks is *Aythya valisneria*. Wild celery is also important to marsh birds and shore birds including rail, plover, sand piper and snipe. Muskrats are also known to graze on it. Beds of wild celery are considered good fish habitat providing shade, shelter and feeding opportunities.

Through the Year: Wild celery overwinters by hardy rhizomes and tubers. Reproduction by seed may occur when conditions are favorable. During the growing season, wild celery stakes out new territory with spreading rhizomes. Flowering occurs midsummer and the podlike fruit is mature by fall. In the fall, vegetative "offsets" break free from rhizomes and float to new locations.

Native

SUBMERSED

Zosterella dubia (Zos-ter-EL-a DEW-bee-a)

(formerly known as Heteranthera dubia)

Water stargrass

Zosterella – (L.) *zostera*: eelgrass (a marine plant with flat leaves) + *ella* – little one; *dubia* – doubtful

The bright yellow flowers of water stargrass swirled on the surface like music box dancers. For a moment, an emerald damselfly hovered and darted near the star-flowered plant.

leaf venation with no prominent midvein

3 x life-size

Description: Water stargrass has slender, freely branched stems that emerge from a buried rhizome. The narrow, alternate leaves (up to 15 cm long, 2-6 mm wide) attach directly to the stem with no leaf stalk and lack a prominent midvein. Yellow, star-shaped flowers are produced individually. The capsular fruit (about 1 cm long) contains 7-30 seeds.

Similar species: This star-flowered plant is often mistaken for a pondweed, yet it is actually part of the pickerelweed family (*Pontederiaceae*). The narrow, alternate leaves of water stargrass can look like a flat-stem pondweed (*Potamogeton zosteriformis*) or small pondweed (*Potamogeton pusillus*) at first glance. However, the leaves of water stargrass lack a definite midvein and when it is in flower, the yellow blossoms clearly separate *Zosterella* from the pondweeds.

Origin & Range: Native; common in Wisconsin; range includes most of U.S.

Habitat: Water stargrass grows in a variety of water depths, from very shallow to several meters deep. It can succeed in a range of sediment types and will tolerate reduced water clarity.

Through the Year: Water stargrass overwinters by hardy rhizomes. There is also some limited reproduction from seed when conditions are favorable. Flowering occurs in midsummer with fruit developing by fall. Stems and leaves die back as the water cools late in the season.

Value in the Aquatic Community: Water stargrass can be a locally important source of food for geese and ducks including northern pintail, blue-winged teal and wood duck. It also offers good cover and foraging opportunities for fish.

life-size

stipule
4 x
life-size

A Closer Look:

The characteristic bright yellow flowers of *Zosterella* are most often produced on plants growing in shallow water. Water stargrass plants in deeper water are usually either sterile or have hidden flowers protected in the bases of submersed leaves (Voss 1972).

Native

Armoracia lacustris (ARE-more-A-cee-a la-CUS-tris)

(also known as Armoracia aquatica or Neobeckia aquatica)

Lake cress

Armoracia – (L.) horseradish; *aquatica* – (L.) of the water

The forested rise overlooking the bay would make a superb site for a vacation home. The couple walked the moist shore looking for a suitable location for the dock. Then they noticed the salad green tufts and trailing stems of the rare lake cress.

Description: This rare chameleon of cresses can assume many forms. Stems of lake cress emerge from a fibrous root-stalk. The submersed leaves are usually finely divided into filament-like segments. Portions of the stem may trail on the water's surface. Leaves produced along the emergent part of the stem are usually lance-shaped to oblong (3-7 cm long) with a sharply lobed or toothed margin. Flowers have four small white petals (6-8 mm long). Inflated, oval fruits (5-8 mm long) may be produced on short stalks. Seeds rarely mature inside the fruit.

Similar species: The emergent leaves of lake cress resemble some of the other aquatic cresses, but the finely divided submersed leaves are quite distinctive. Lake cress is part of the mustard family. It is closely related to water cress, but much less common.

Origin & Range: Native; lake cress has only been found at a few scattered locations in eastern and northern Wisconsin, including sites in Lake Superior estuaries. Historic range has included 26 states in the U.S., but in recent years it has disappeared from many sites (Les 1994). It is considered rare in most of its range in the U.S. and is listed as **Endangered in Wisconsin**.

Habitat: Lake cress can be found in quiet water or on moist shorelines. It will grow in water several meters deep.

Through the Year: Lake cress over-winters by hardy rootstalks. New stems emerge in spring. Growth may remain entirely submersed, or portions of the stem may reach the surface and trail on the water. Most reproduction occurs vegetatively. Arching stems can take root in the sediment and send up new shoots; sections of leaves and stems break off and float to new locations, and leaf fragments can generate new "plantlets."

Value in Aquatic Community: Lake cress is rare. Where is does occur, it can provide habitat for invertebrates and shade and foraging opportunities for fish.

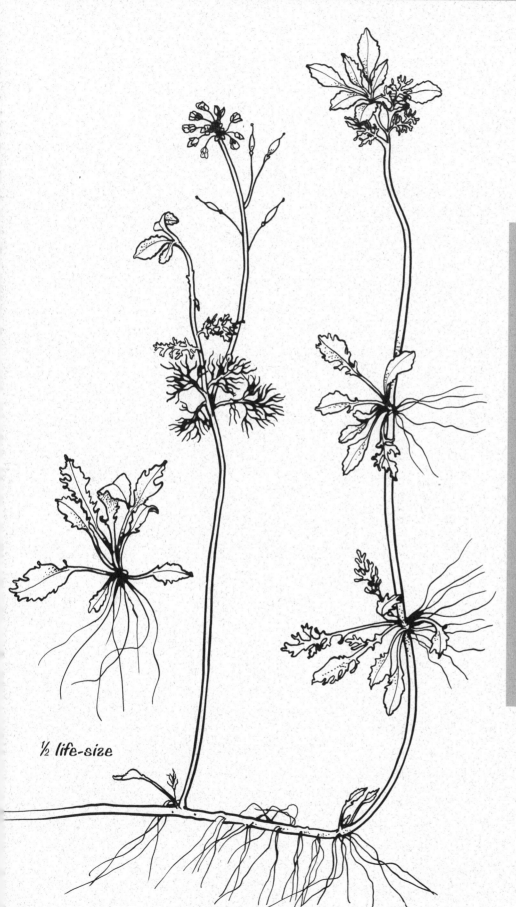

½ life-size

A Closer Look:

All of the water loving cresses (*Armoracia*, *Cardamine*, *Nasturtium*, *Neobeckia* and *Rorippa*) have extremely variable appearance. This has made it difficult to decide relationships among the plants. Because lake cress is disappearing from much of its range, it has become important to understand its affiliation to the other cresses.

DNA studies have revealed a close evolutionary relationship between the aquatic cresses. However, DNA sequence data shows lake cress is unique enough to be classified as a distinct genus, *Neobeckia aquatica*, rather than being combined with *Armoracia* (horseradish) or *Rorippa* (yellow cress) (Les 1994).

R a r e

Bidens beckii (BYE-denz BECK-ee-i)

(formerly known as Megalodonta beckii)

Water marigold

Bidens – (L.) two-toothed;
beckii – named for its discoverer, Lewis Caleb Beck (1798-1853)

Water marigold has a split personality. Below the water's surface, the leaves are divided into a fine lacework of green filigree. Above the surface, the same stem bears leaves that look like they belong on a sunflower. Each leaf is perfectly adapted to its surrounding, taking advantage of life in the air or life in the water.

Description: The stems of water marigold emerge from a buried root-stalk. The submersed leaves are finely cut into many thread-like divisions. Often only the underwater portion of the stem is present. When an aerial portion develops, the emersed leaves are lance-shaped, have a toothed margin and attach directly to the stem. When flowering occurs, a yellow daisy-like bloom (2-2.5 cm wide) develops on a sturdy stalk above the water surface. The central portion of the flower produces narrow fruits, each with 3-6 long, barbed bristles.

Similar species: When only the submersed portion of water marigold is present, it can be confused with coontail (*Ceratophyllum demersum*) or water crowfoot (*Ranunculus* sp.). Coontail has leaves that are stiff and toothed along the margin, while those of water marigold are flexible and lack teeth. Water marigold differs from water crowfoot in leaf position. The leaves of water crowfoot are alternate on the stem, while those of water marigold appear to be whorled. When water marigold is blooming, it won't be confused with any other submersed plant.

Origin & Range: Native; found primarily in northern and eastern Wisconsin; range includes the eastern U.S. and scattered locations in the west.

Habitat: Water marigold is usually found growing in soft sediment in clear water lakes. It will grow from ankle-deep water up to almost 3 meters deep.

Through the Year: Water marigold overwinters by hardy rootstalks and rhizomes. When conditions are right, it can reproduce from seed. Flowering may occur in midsummer and fruit is mature by late summer.

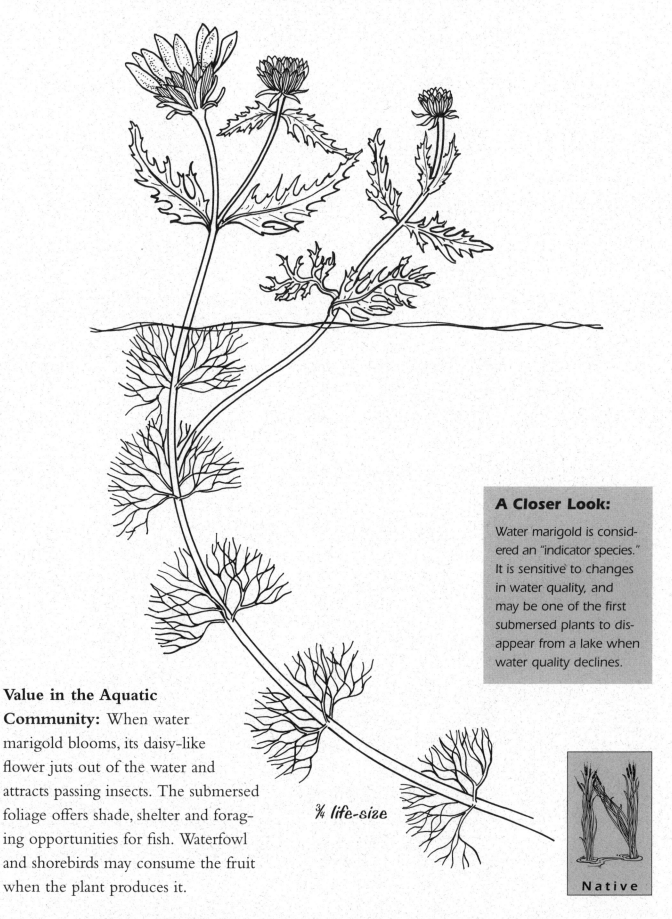

¾ life-size

A Closer Look:

Water marigold is considered an "indicator species." It is sensitive to changes in water quality, and may be one of the first submersed plants to disappear from a lake when water quality declines.

Value in the Aquatic Community: When water marigold blooms, its daisy-like flower juts out of the water and attracts passing insects. The submersed foliage offers shade, shelter and foraging opportunities for fish. Waterfowl and shorebirds may consume the fruit when the plant produces it.

Native

Ceratophyllum demersum (cer-at-oh-FILL-um de-MER-sum)

Coontail, hornwort

Ceratophyllum (Gk.) *ceras:* a horn + *phyllon:* leaf; *demersum* − (L.) submerged

Taking a winter sample of coontail is an eye-opening experience. Pulling the long, bushy strands up through a hole in the ice reveals a lively scene where the fine leaves are still green and hopping with invertebrates.

Description: Coontail has long, trailing stems that lack true roots. However, the plant may be loosely anchored to the sediment by pale modified leaves. The leaves are stiff and arranged in whorls of 5-12 at a node. Each leaf (1-3 cm long) is forked once or twice. The leaf divisions have teeth along the margins that are tipped with a small spine. Whorls of leaves are usually more closely spaced near the ends of branches, creating the raccoon tail appearance.

Flowers are tiny and hidden in the axils of leaves. Male and female flowers are on separate plants. The stamen of the male plants float to the surface at maturity and discharge pollen. The pollen sinks down through the water and may or may not land on the tiny female flowers, tucked in the leaf axils. Fruit is rarely produced, partly because of this unpredictable method of pollination. When fruit does develop, it is a nut-like achene with two spines at the base and one on top (the persistent style). *(continued)*

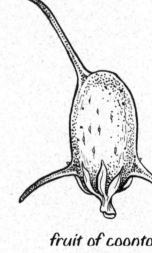

*fruit of coontail
6 x life-size*

*tooth
with spine
3 x life-size*

COONTAIL

½ life-size

A Closer Look:

Coontail has the capacity to grow at nuisance levels. Management strategies are often designed to reduce the amount of coontail present in a water body. However, reduction and not elimination should be the goal, because coontail does offer good habitat.

The ability of *Ceratophyllum* to draw nutrients from the water has lead to some creative uses of the plant. Inventive pond managers have designed "coontail pods" – permeable containers filled with coontail. These pods are placed in small ponds to help reduce phosphorus levels and thereby inhibit algae growth.

Native

Ceratophyllum demersum (continued)

Similar species: There is one other specie of *Ceratophyllum* in Wisconsin:

Spiny hornwort (*Ceratophyllum echinatum*) is usually only found in low pH, soft-water lakes and ponds. It has recently been listed as a **Species of Special Concern** in Wisconsin.

The leaves of coontail are usually twice forked and have a distinctly toothed margin; those of spiny hornwort are forked 3-4 times and usually lack teeth, although small spines are present. The fruit of coontail has two basal spines. The fruit of spiny hornwort, by contrast, has several spines of differing lengths around its margin and a rough surface.

fruit of spiny hornwort
6 x life-size

Origin and Range: Native; common throughout Wisconsin; range includes most of the U.S.

Habitat: Coontail has a tolerance for low light conditions and will grow in water several meters deep. Because it is not rooted, it can drift between depth zones.

Through the Year: A tolerance for cool water and low light conditions allows coontail to overwinter as an evergreen plant, continuing photosynthesis at a reduced rate under the ice. Vigorous growth resumes in spring. New plants are formed primarily by stem fragmentation, because seeds rarely develop.

Value in the Aquatic Community: The stiff whorls of leaves offer prime habitat for a host of critters, particularly during the winter when many other plants are reduced to roots and rhizomes. Look for scuds (*Gammarus* spp. and *Hyalella* spp.) that are often found on coontail in this cool water. Both foliage and fruit of coontail are grazed by waterfowl including black duck, bufflehead, canvasback, gadwall, mallard, northern pintail, redhead, scaup, blue-winged teal, green-winged teal and wood duck. Bushy stems of coontail harbor many invertebrates and provide important shelter and foraging opportunities for fish.

Similar Species:
SPINY HORNWORT

small spine
on leaf margin
10 x life-size

life-size

3 x life-size

each leaf forks
3-4 times

Native

SUBMERSED

Myriophyllum farwellii (MIR-ee-o-FILL-um far-WELL-ee-i)

Farwell's water milfoil

Myriophyllum – (Gk.) *myrio:* many + *phyllon:* leaf;
farwellii – named for its discoverer, Oliver A. Farwell (1868-1944)

Buoyed by the water, the stems of Farwell's water milfoil resembled a fine feather boa. When pulled from the water, the delicate stems collapsed in a limp green mass.

Description: While the more common water milfoils in this area have whorled leaves, Farwell's has closely spaced, scattered leaves that create a bushy appearance. The delicate stems (less than 1 m long) emerge from the rootstalk. Closely spaced, scattered leaves cover the stem. Each leaf (1-3 cm long) has a short stalk and is divided like a feather into pairs of slender leaflets (less than 14 pairs). Flowers develop in the axils of submersed leaves. Each flower produces a hard four-parted fruit (2-2.5 mm long). Each section of the fruit has a pair of bumpy, longitudinal ridges.

Similar Species: Farwell's water milfoil looks similar to two other water milfoils found in this region. It can be separated from *Myriophyllum humile,* which occasionally occurs here, by the fruit. The fruit of *M. humile* has a smooth surface, while Farwell's has ridges. It can be distinguished from various-leaved water milfoil (*M. heterophyllum*) by the position of the fruit. Various-leaved water milfoil produces fruit in terminal spikes, while Farwell's bears fruit nestled in the leaf axils of submersed leaves. Late in the season, Farwell's produces winter buds while the other two species do not.

Origin & Range: Native; found at scattered locations, primarily in northern Wisconsin; Farwell's water milfoil is listed as a **Special Concern** species in Wisconsin; range includes northeastern U.S.

Habitat: Farwell's water milfoil is usually found growing in fine sediment in soft-water lakes. It grows from shallow zones to water 2 meters deep.

Through the Year: New shoots emerge in spring from overwintering rootstalks or winter buds. Flowers and fruit develop in the leaf axils as the growing season progresses. Winter buds of tightly compressed leaves are formed late in the season.

Value in the Aquatic Community: The bushy form of Farwell's stems traps detritus that provides food and affords favorable habitat for invertebrates. Both foliage and fruit are sometimes grazed by waterfowl. Beds of Farwell's water milfoil provide shelter and foraging opportunities for fish.

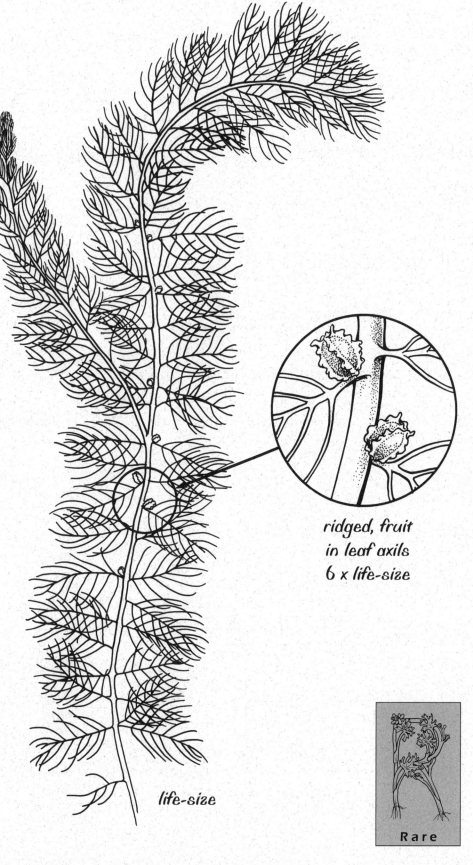

ridged, fruit
in leaf axils
6 x life-size

life-size

A Closer Look:

Farwell's water milfoil is named in honor of O. A. Farwell, who collected many plants in Michigan at the turn of the century. The plant used to name the species, called the type specimen of *Myriophyllum farwellii*, was collected by Farwell from a small pond that was only 0.5 meter deep (Voss 1985).

Rare

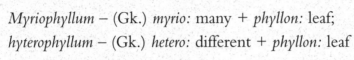

Myriophyllum heterophyllum
(MIR-ee-o-FILL-um HET-er-o-FILL-um)

Various-leaved water milfoil

Myriophyllum – (Gk.) *myrio:* many + *phyllon:* leaf;
hyterophyllum – (Gk.) *hetero:* different + *phyllon:* leaf

*The sunlight cuts a swath through the feathery leaves of the
water milfoil. A school of minnows gleams golden as they swim
from the shadows through the sunrays, like actors moving
through the spotlight on a stage.*

Description: Various-leaved water milfoil has stems that emerge from a hardy rootstalk. The leaves (2-4 cm long) are divided like a feather, with a short stalk and about 7-10 pairs of thread-like leaflets. Most of the leaves are arranged in whorls (4-6 leaves in a whorl), but some are scattered on the stem. The whorls are closely spaced, usually less than 10 mm apart.

Flowers are clustered in a spike that sticks up out of the water. The flowers develop in the axils of bracts (4–18 mm long) that are blade-shaped with a deeply toothed margin. The middle and upper bracts on the spike are much longer than the flowers and fruit. Flowers are small with a red to pink tint. The fruit (1-1.5 mm) is hard and looks a bit like the top of a clove. It has four sections, each with a couple of low ridges on the back and a curved beak. No winter buds are formed.

Similar species: There are seven other water milfoils that occur in this region, some more commonly than others. Dwarf water milfoil is easy to separate because it is small with greatly reduced leaves (see *Myriophyllum tenellum*). There are two species with leaves that are mostly scattered and have fruit in the axils of submersed leaves (see discussion at *Myriophyllum farwellii*).

Various-leaved water milfoil could be confused with the four other species of water milfoil that have whorled leaves:

Northern water milfoil (*M. sibiricum*) may have the same number of leaf divisions as various-leaved water milfoil. However, it differs in having flower bracts that are short and either entire or slightly indented. It also has cylindrical winter buds that are formed late in the growing season. *(continued)*

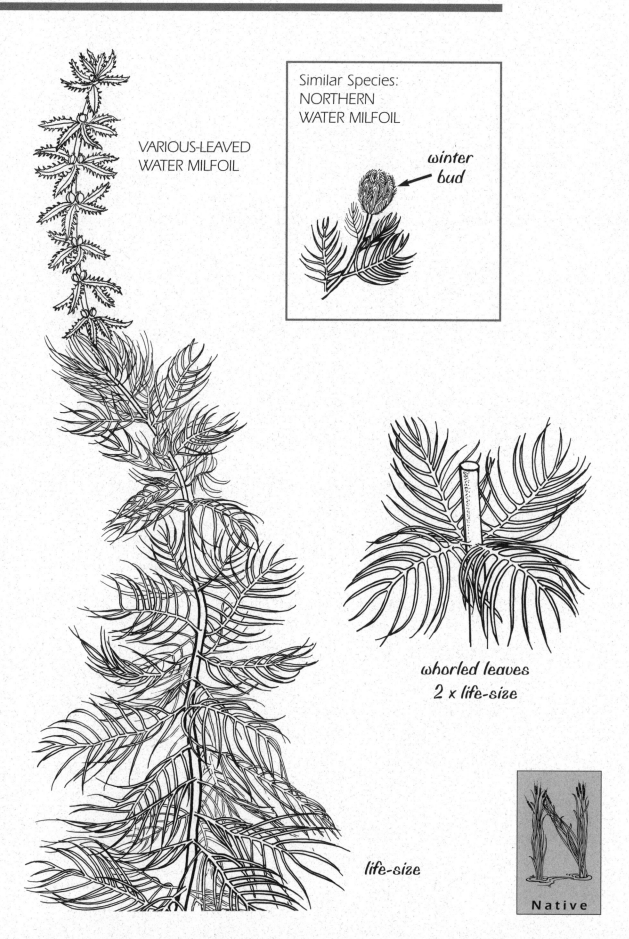

VARIOUS-LEAVED
WATER MILFOIL

Similar Species:
NORTHERN
WATER MILFOIL

winter bud

whorled leaves
2 x life-size

life-size

Native

Eurasian water milfoil (*Myriophyllum spicatum*) can be distinguished by the number of leaf divisions. It usually has more than 14 pairs of leaf divisions, while various-leaved water milfoil has less than 14 (usually 7-10 pairs). No winter buds are formed.

Alternate-flowered water milfoil (*Myriophyllum alterniflorum*) is distinct because it has smaller leaves than the other whorled species (usually less than 1 cm long). The flower spike is also different, having alternate flowers and bracts rather than whorled.

Whorled water milfoil (*M. verticillatum*) may have the same number of leaf divisions as various-leaved water milfoil. However, it differs in having leaves that attach directly (no leaf stalk) to a greenish-brown stem and flower bracts that are lobed. Club-shaped winter buds are formed late in the growing season.

A few other guidelines help separate various-leaved water milfoil from the other whorled species. If the space between most of the whorls is more than 10 mm, it probably isn't various-leaved water milfoil. But if the foliage is dense, with tightly spaced whorls and 5-6 leaves per whorl, then it probably is various-leaved water

milfoil. Various-leaved water milfoil also has a tendency to produce some alternate leaves, while the other whorled species don't (Voss 1985).

Origin & Range: Native; scattered locations throughout Wisconsin; range includes most of U.S.

Habitat: Various-leaved water milfoil grows on a variety of sediments in water up to 5 meters deep. It is often found in large, dense stands.

Through the Year: Various-leaved water milfoil overwinters by hardy root-stalks and rhizomes. It does not produce winter buds. New growth begins in spring and plants can reach the surface by early to midsummer. Flowering may or may not occur, depending on conditions. Stem fragments often break off and colonize new areas during the growing season.

Value in the Aquatic Community: The fruit and foliage of various-leaved water milfoil are consumed by a variety of waterfowl. The extensive foliage traps detritus for food and provides invertebrate habitat. Beds of various-leaved water milfoil offer shade, shelter and foraging opportunities for fish.

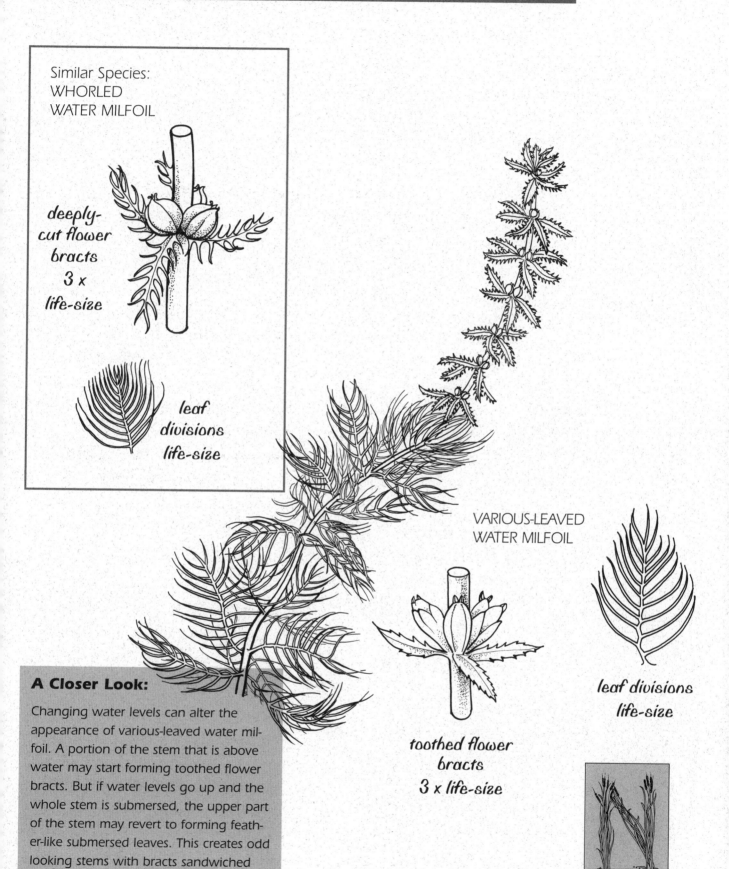

Similar Species:
WHORLED
WATER MILFOIL

deeply-cut flower bracts
3 x life-size

leaf divisions life-size

VARIOUS-LEAVED
WATER MILFOIL

leaf divisions life-size

toothed flower bracts
3 x life-size

A Closer Look:

Changing water levels can alter the appearance of various-leaved water milfoil. A portion of the stem that is above water may start forming toothed flower bracts. But if water levels go up and the whole stem is submersed, the upper part of the stem may revert to forming feather-like submersed leaves. This creates odd looking stems with bracts sandwiched between normal looking whorls of leaves.

Native

SUBMERSED

Myriophyllum sibiricum (MIR-ee-o-FILL-um si-BIR-i-cum)

(formerly known as Myriophyllum exalbescens)

Northern water milfoil, spiked water milfoil

Myriophyllum – (Gk.) *myrio:* many + *phyllon:* leaf; *sibiricum* – of Siberia

The feathery stems of northern water milfoil rose from the soft bottom like spires on a gothic cathedral. Below the green architecture, mortal combat was under way. The crayfish backed away slowly, its claws raised in a stern warning. The bass moved in, mindful of the armored pincers.

A Closer Look:

Formation of winter buds is triggered by changes in light and temperature. The number of buds formed varies from year to year depending on weather conditions.

Description: Northern water milfoil has light-colored stems that emerge from rootstalks and rhizomes. If a stem has sprouted from a winter bud, it will be U-shaped at the base. Stems are sparingly branched and fairly erect in the water. Leaves (1-5 cm long) are divided like a feather, with a short stalk and about 5-12 pairs of thread-like leaflets. The lower leaflet pairs are longer than the upper ones, creating a Christmas tree shape. The leaves are arranged in whorls (4–5 leaves per whorl).

The flower spike sticks up out of the water with whorls of red-tinted flowers in the axils of short bracts. Lower bracts may be slightly indented, while the upper bracts are usually entire. The fruit (about 2 mm long) is four-parted with a smooth to slightly roughened surface.

Similar species: There are seven other species of water milfoil found in this region. Northern water milfoil most closely resembles three other species with whorled leaves: various-leaved water milfoil (*M. heterophyllum*), **whorled water milfoil** (*M. verticillatum*) and Eurasian water milfoil (*M. spicatum*). (See discussion under *Myriophyllum heterophyllum*.)

Origin & Range: Native; common throughout Wisconsin, particularly in the eastern and northern sections; range includes northern U.S. and parts of the west.

Habitat: Northern water milfoil is usually found growing in soft sediment in fairly clear-water lakes. It can grow from shallow zones to depths over 4 meters. Northern water milfoil is sensitive to reduced water clarity and has declined in lakes that are becoming eutrophic.

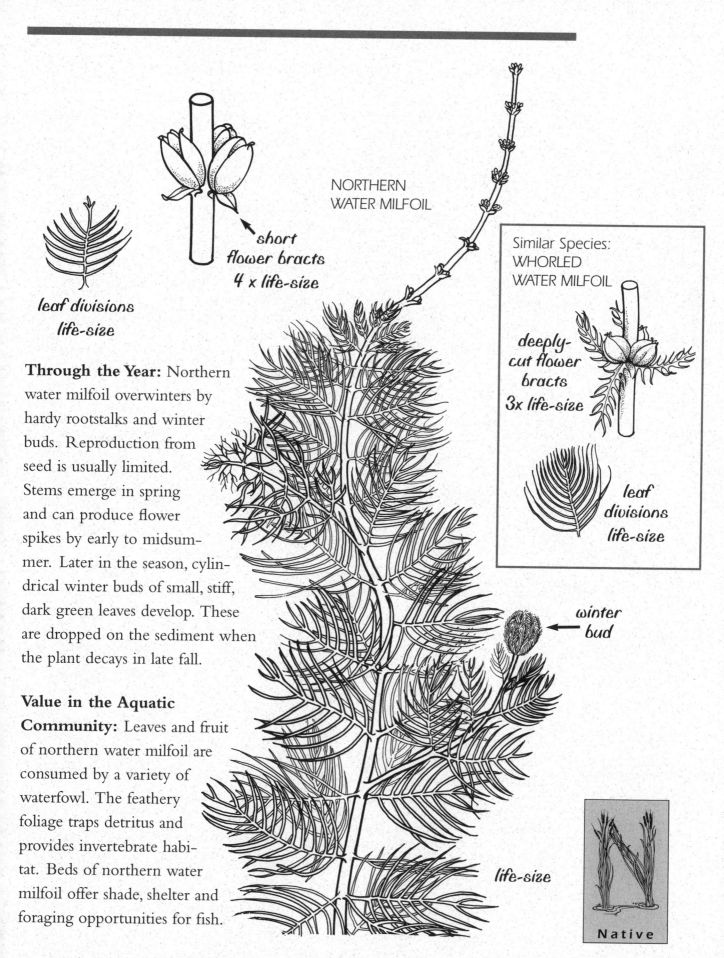

leaf divisions
life-size

NORTHERN
WATER MILFOIL

short
flower bracts
4 x life-size

Similar Species:
WHORLED
WATER MILFOIL

deeply-
cut flower
bracts
3x life-size

leaf
divisions
life-size

Through the Year: Northern water milfoil overwinters by hardy rootstalks and winter buds. Reproduction from seed is usually limited. Stems emerge in spring and can produce flower spikes by early to midsummer. Later in the season, cylindrical winter buds of small, stiff, dark green leaves develop. These are dropped on the sediment when the plant decays in late fall.

Value in the Aquatic Community: Leaves and fruit of northern water milfoil are consumed by a variety of waterfowl. The feathery foliage traps detritus and provides invertebrate habitat. Beds of northern water milfoil offer shade, shelter and foraging opportunities for fish.

winter
bud

life-size

Native

Myriophyllum spicatum (MIR-ee-o-FILL-um spi-KAY-tum)

Eurasian water milfoil

Myriophyllum – (Gk.) *myrio:* many + *phyllon:* leaf; *spicatum* – (L.) pointed, spiked

Eurasian water milfoil has made quite a name for itself. Its likeness shows up in newspapers, on bumper stickers, billboards and public service announcements. A plant that can grow as swiftly and tenaciously as this one gets people's attention. You could say Eurasian water milfoil is a real "growth business."

Description: Eurasian water milfoil has long, spaghetti-like stems, sometimes 2 or more meters in length, that emerge from roots and rhizomes. Stems often branch repeatedly at the water's surface, creating a canopy of floating stems and foliage. Leaves are divided like a feather, with a short stalk and about 14-20 pairs of thread-like leaflets. The leaf divisions are all about the same length and closely spaced, resembling the bones on a fish spine. Leaves are in whorls of 4-5, and can be widely spaced (1-3 cm or more). The flower spike sticks out of the water with whorls of flowers in the axils of short bracts. The fruit (2-3 mm) has four parts with a smooth to slightly roughened surface.

*leaf
divisions
life-size*

Similar species: There are seven other species of water milfoil in this region, which are all native (see discussion under *Myriophyllum heterophyllum*). Eurasian water milfoil most closely resembles northern water milfoil (*Myriophyllum sibiricum*). The most reliable way to distinguish between them is by the number of leaf divisions. Eurasian water milfoil usually has more than 14 pairs of leaflets, whereas northern water milfoil has less than 14 (usually 5-12). The presence or absence of winter buds in late summer is also a helpful characteristic. Northern water milfoil produces winter buds, but Eurasian water milfoil does not.

Although individual plants of these two species may look similar, their growth form is quite different. Northern water milfoil doesn't typically form a branched canopy at the water's surface and it grows in a more controlled manner with slower stem growth and less fragmentation.

Origin & Range: Exotic, originated in Europe and Asia; distribution in Wisconsin is primarily in the south, but spreading north; range includes most of U.S.

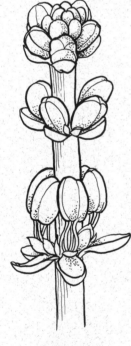

*flower
spike
6 x life-size*

Habitat: Eurasian water milfoil is usually found in water 1 to 4 meters deep. It can grow in a variety of sediments, but is most productive in fine textured, inorganic sediment (Barko and Smith 1986). Low light and high water temperatures promote canopy formation.

Through the Year: Eurasian water milfoil does not produce winter buds or tubers. Some shoots may overwinter and others develop from sprouts on the rootstalk. Growth can begin early in the spring when water temperatures are still cool (about 59°F). Plants growing in shallow water can reach the surface within a few weeks, while those growing in deeper water may not reach the surface until late in the growing season. Flower stalks don't develop until the stems reach the surface. After flowering and fruit production, portions of the stems break apart in fragments. These fragments can float to new locations and take root. If the first flowering cycle occurs early in the growing season, it may be repeated again in the fall (Smith and Barko 1990).

Value in the Aquatic Community: Waterfowl graze on fruit and foliage to a limited extent. Milfoil beds provide invertebrate habitat, but studies have shown mixed stands of pondweeds and wild celery have higher diversity and numbers of invertebrates (Engel 1990).

A Closer Look:

Eurasian water milfoil is an exotic species introduced to the United States from its native range in Europe and Asia. Its fast growing shoots and extensive canopy formation can obstruct recreation and navigation. The ability to grow in cool water gives it a quick start in the spring. Eurasian water milfoil often crowns and shades native plants, giving it a competitive advantage.

Eurasian water milfoil has been the target of many management strategies ranging from harvesting to herbicides. There has recently been some evidence that a native weevil (*Euhrychiopsis lecontei*) may provide a biological control. This tiny aquatic weevil has been associated with some natural declines of Eurasian water milfoil (including Brownington Pond, Vermont and Fish Lake, Wisconsin). In test plots at milfoil infested lakes, the weevil has been shown to reduce milfoil growth and limit canopy formation. The possibility of using a native weevil for biocontrol shows promise (Sheldon and Creed 1995).

life-size

E

Exotic

Ranunculus longirostris (rah-NUN-cue-les lon-gi-ROS-tris)

Stiff water crowfoot, white water crowfoot, white water buttercup

Ranunculus – (L.) little frog; *longirostris* – (L.) long-beaked (referring to fruit)

The water crowfoot is in bloom. Its spreading underwater stems send up hundreds of aerial flower stalks. Each stalk is tipped with a delicate white blossom that appears to float in the air just above the water's surface, creating the illusion of a cloud hovering over the still water.

Description: Stiff water crowfoot has long, branched stems that emerge from both trailing runners and buried rhizomes. The leaves (1-2 cm long) are finely cut into thread-like divisions and either attach directly to the stem or have a very short leaf stalk. Leaves emerge along the stem in an alternate arrangement and are stiff enough to hold their shape when lifted out of the water. White, five-petaled flowers (1-1.5 cm wide) are produced on stalks of varying lengths, just above the water's surface. As the flowers develop into fruit, the stalks curve back into the water. A cluster of 15-25 nutlets is produced, each with a slender beak (0.7-1.5 mm).

Similar species: Two white-flowered water crowfoot species are common in our region: stiff water crowfoot (*Ranunculus longirostris*) and **white water crowfoot** (*Ranunculus trichophyllus*). Both have leaves divided into thread-like divisions, but the leaves of stiff water crowfoot are stiff and have almost no leaf stalk, while those of white water crowfoot are limp and have a noticeable stalk between the stem and divided portion of the leaf. The fruits are also different. The nutlets of stiff water crowfoot have long, slender beaks (0.7-1.5 mm); those of white water crowfoot have tiny beaks (only 0.2-0.5 mm).

Yellow water crowfoot species (including *Ranunculus flabellaris*) could be mistaken for white water crowfoot when they are not in flower. However, leaf divisions of the yellow water crowfoot species are always flat rather than thread-like.

Origin & Range: Native; scattered locations throughout Wisconsin; range includes most of U.S.

Habitat: Stiff water crowfoot is found in both lakes and streams with higher alkalinity, usually in less than 2 meters of water.

A Closer Look:

Ranunculus spp. have been used in the past for a variety of medicinal purposes, including treatment of rheumatism and arthritis. They contain some potent compounds including one unique to the genus – *ranunculin*.

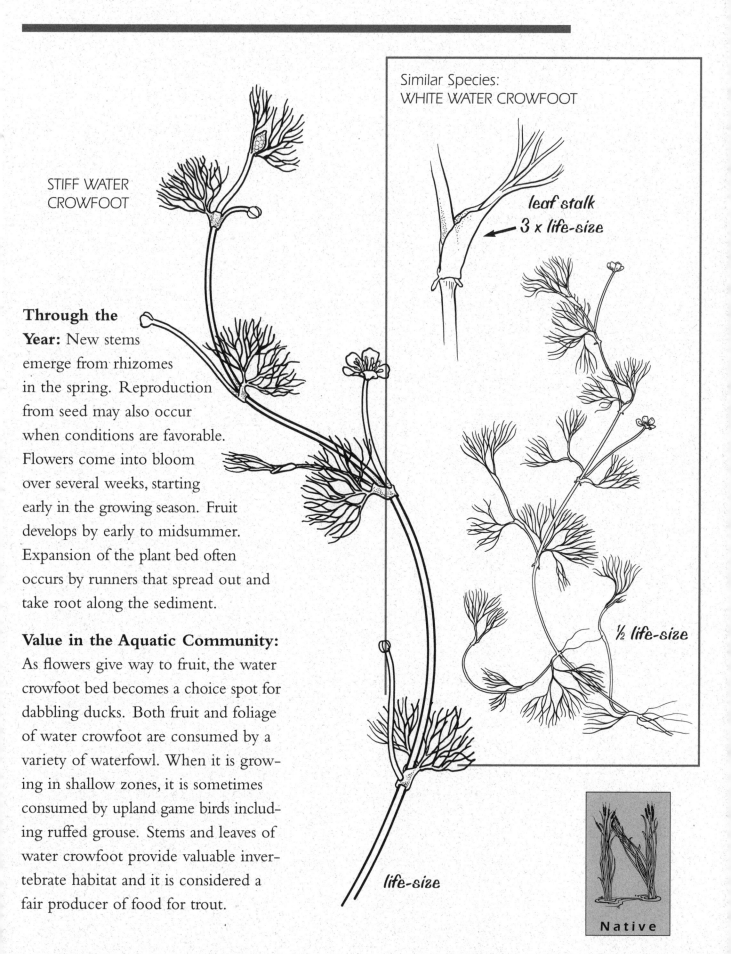

STIFF WATER
CROWFOOT

Similar Species:
WHITE WATER CROWFOOT

leaf stalk
← *3 x life-size*

½ life-size

Through the Year: New stems emerge from rhizomes in the spring. Reproduction from seed may also occur when conditions are favorable. Flowers come into bloom over several weeks, starting early in the growing season. Fruit develops by early to midsummer. Expansion of the plant bed often occurs by runners that spread out and take root along the sediment.

Value in the Aquatic Community: As flowers give way to fruit, the water crowfoot bed becomes a choice spot for dabbling ducks. Both fruit and foliage of water crowfoot are consumed by a variety of waterfowl. When it is growing in shallow zones, it is sometimes consumed by upland game birds including ruffed grouse. Stems and leaves of water crowfoot provide valuable invertebrate habitat and it is considered a fair producer of food for trout.

life-size

Native

Utricularia gibba (u–TRICK–u–LAIR–ee–a GIB–ba)

Creeping bladderwort, humped bladderwort, cone-spur bladderwort

Utricularia – (L.) *utriculus:* a small bag or bladder; *gibba* – (L.) humped

The woman pulled the kayak up on the muddy shoreline of the pond to observe the thread-like stems of creeping bladderwort. They resembled bits of a torn hairnet as they washed up on the exposed mudflats and draped over the driftwood along the shore.

**bladder
25 x
life-size**

Description: The delicate, free-floating stems of creeping bladderwort are usually only about 10 cm long. They often form tangled mats in shallow water or stranded on shore. Side branches (about 5 mm long) fork once or twice and have a few scattered bladders. Flower stalks (5-10 cm long) produce 1-3 yellow, snapdragon-like flowers (only 5-6 mm long). The lower lip of the flower has a short spur. The fruit is a small pod with several seeds.

Similar species: The fine stems are sometimes mistaken for algae. A close look reveals the characteristic bladders. (See *Utricularia vulgaris* for description of traps.) When creeping bladderwort is blooming on shore, it could be mistaken for horned bladderwort.

Horned bladderwort (*Utricularia cornuta*) grows anchored in the shoreland soil. It has finely branched buried stems with minute bladders. It can usually be spotted only when it is blooming. The flower stalk (10-25 cm tall) has 1-6 yellow flowers that are larger than those of creeping bladderwort. The yellow flowers also have a prominent spur (7-14 mm long) on the lower lip.

Origin & Range: Native; uncommon in Wisconsin; range includes eastern U.S. and the western coast.

Habitat: Creeping bladderwort is found from moist shorelines to water several meters deep. It is free-floating and moves between depth zones. It is often found in quiet waters associated with bogs.

Through the Year: Creeping bladderwort may survive the winter as stem fragments resting on the sediment. Active growth starts as the water warms in spring. Flower stalks develop in early summer and blooms can last for several weeks.

Value in Aquatic Community: Mats of creeping bladderwort offer cover and foraging opportunities for fish.

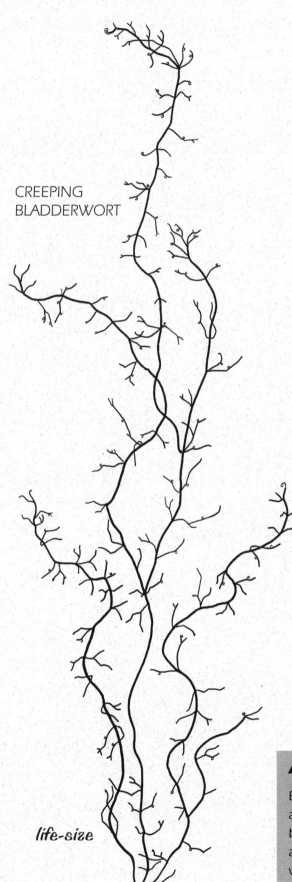

CREEPING
BLADDERWORT

life-size

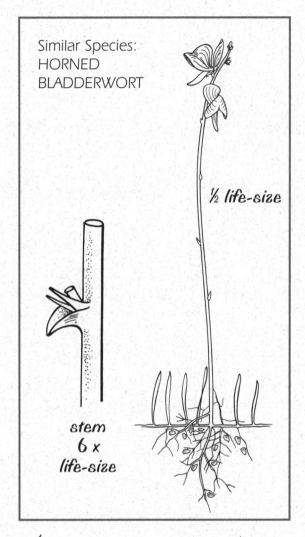

½ life-size

stem
6 x
life-size

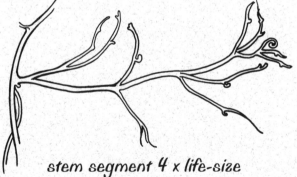

stem segment 4 x life-size

A Closer Look:

Bladderworts make an interesting aquarium or water garden plant. Creeping bladderwort can be grown in a tray with a 1:1 mixture of sphagnum moss and water (Whitley et al. 1990).

Native

Utricularia purpurea (u-TRICK-u-LAIR-ee-a PUR-pur-ee-a)

Large purple bladderwort

Utricularia – (L.) *utriculus:* a small bag or bladder; *purpurea* – (L.) purple

The diver swam along the reef-like edge of a dense wall of large purple bladderwort. She gently cradled a strand of its curly green tendrils in her hand. What at first glance had appeared to be tiny beads were actually miniscule traps.

Description: The free-floating stems (up to 1 m long) of large purple bladderwort produce whorls of filament-like branches. The tips of these side branches have small bladders used for capturing prey. Flower stalks emerge above the water's surface. Several purple, snapdragon-like flowers are produced on each stalk. The lower lip of the flower is three-lobed with a yellow spot.

Similar species: Large purple bladderwort can be separated from other bladderworts in this region by the whorled branching pattern (see descriptions of other species at *Utricularia vulgaris*).

Origin & Range: Native; scattered locations in Wisconsin – listed as a **Special Concern** species in Wisconsin; range includes eastern U.S.

Habitat: Large purple bladderwort is usually found in quiet waters of softwater, low pH lakes. It can grow from shallow zones to water several meters deep.

Through the Year: Large purple bladderwort survives the winter by stem fragments or winter buds. Active growth begins as water warms in the spring. Flowering occurs over several weeks during the middle of the growing season. Winter buds are formed on side branches later in the summer.

Value in the Aquatic Community: Masses of large purple bladderwort offer invertebrate habitat and foraging opportunities for fish.

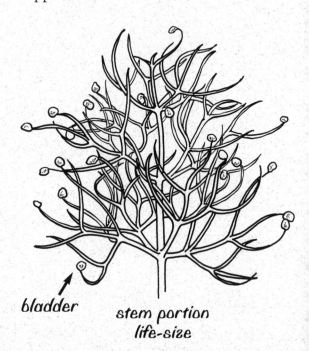

bladder

stem portion life-size

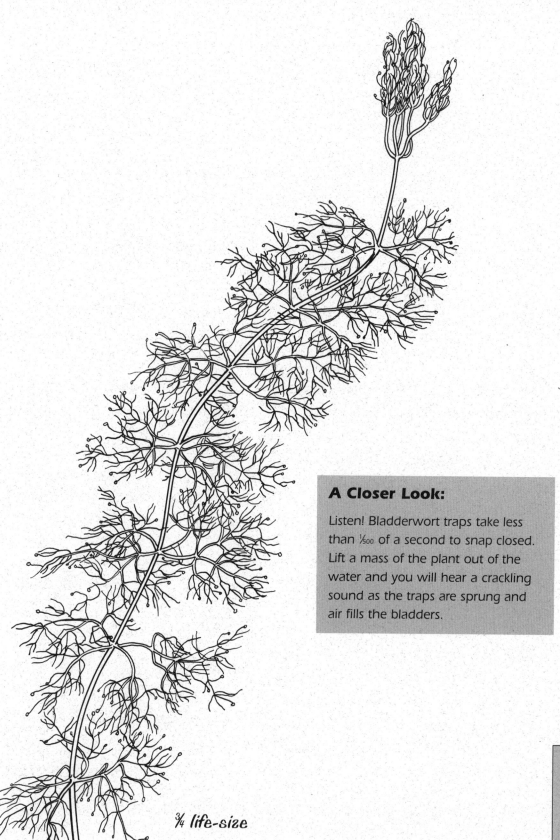

¾ life-size

A Closer Look:

Listen! Bladderwort traps take less than ⅟₆₀₀ of a second to snap closed. Lift a mass of the plant out of the water and you will hear a crackling sound as the traps are sprung and air fills the bladders.

Rare

Utricularia resupinata
(u–TRICK–u–LAIR–ee–a re–SUE–pin–a–ta)

Small purple bladderwort

Utricularia – (L.) *utriculus:* a small bag or bladder; *resupinata* – (L.) bent back (referring to tilted flower position)

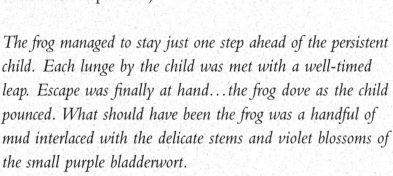

The frog managed to stay just one step ahead of the persistent child. Each lunge by the child was met with a well-timed leap. Escape was finally at hand…the frog dove as the child pounced. What should have been the frog was a handful of mud interlaced with the delicate stems and violet blossoms of the small purple bladderwort.

Description: You may not be able to see small purple bladderwort even as you stand directly over it. The best chance of spotting it is when it is in bloom or occasionally when the stems are washed out into shallow water. The fine stems of small purple bladderwort spread horizontally under the soil surface with some scattered bladders (see *Utricularia vulgaris* for description of bladders). Narrow, grass–like blades (up to 3 cm tall) may be produced along the stems. The flower stalk (2-10 cm tall), as fine as human hair, emerges from the soil and bears a single, violet blossom that is tipped so that it faces upward. Below the flower is a tubular notched bract.

Similar species: The other small bladderworts in this region have yellow flowers (see *Utricularia gibba*). When it is not blooming, the small grass–like blades of small purple bladderwort could be mistaken for a sprouting spikerush or grass.

Origin & Range: Native; scattered locations in Wisconsin – listed as a **Special Concern** species in Wisconsin; range includes the eastern U.S.

Habitat: Small purple bladderwort is usually found on muddy shorelines or shallow water of undisturbed lakes and ponds.

Through the Year: This tiny bladder–wort overwinters as buried stems. Emergent blades and flower stalks may be produced by midsummer, but growth varies dramatically from season to season.

Value in Aquatic Community: The delicate stems of purple bladderwort can offer some limited cover for invertebrates in shallow water.

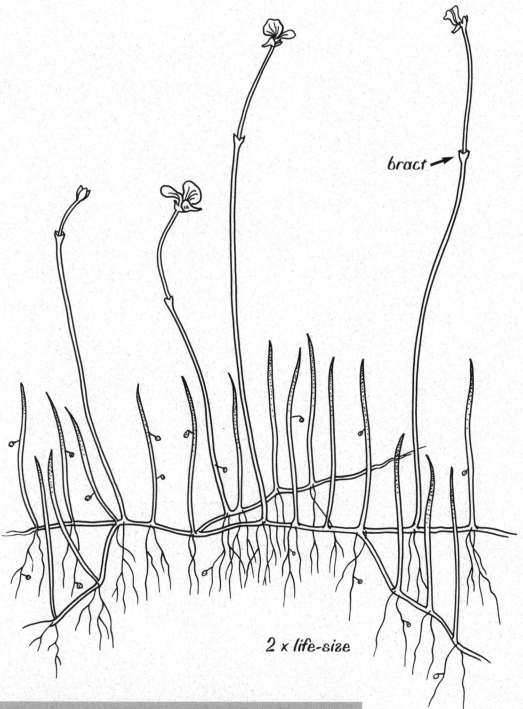

bract →

2 x life-size

A Closer Look:

The genus name *Utricularia* is derived from the Latin word "utriculus," which means small bag. This refers to the sack-like traps. There are over 300 species of bladderwort found throughout the world and some grow in remarkable places. In tropical regions, they can be found growing in the water cups of plants growing on tree trunks.

Rare

SUBMERSED

Utricularia vulgaris (u-TRICK-u-LAIR-ee-a vul-GAR-es)

Common bladderwort, great bladderwort

Utricularia - (L.) *utriculus:* a small bag or bladder; *vulgaris* - (L.) common

Trailing strands of common bladderwort drift in a quiet backwater. Dozens of snapdragon-like flowers hover over the water like a yellow mist. Beneath the surface, stems peppered with bladder-like traps capture small prey.

Description: Common bladderwort has floating stems that can reach 2-3 meters in length. Along the stem are leaf-like branches that are finely divided. The divisions are filament-like, have no midrib, and fork 3-7 times. Scattered on these branches are the bladders that trap prey. Young bladders are transparent and green tinted, but they become dark brown to black as they age. The branches also have fine spines (spicules) scattered along their margins.

Yellow, two-lipped flowers are produced on stalks that protrude above the water surface. There may be 4-20 flowers per stalk. The upper lip of the flower creates an awning over the saclike pouch and sickle-shaped spur of the lower lip. The plant is branched in several directions at the base of the flower stalk. This creates a stable base that keeps the top-heavy flower stalk from capsizing.

Similar species: There are several other free-floating bladderworts you may encounter in this region. Three of these species resemble common bladderwort.

Twin-stemmed bladderwort

(*Utricularia geminiscapa*) looks like a smaller version of common bladderwort. Key differences include small flowers, the presence of some non-opening (cleistogamous) flowers, and spines that are only on the tips of leaf divisions. Twin-stemmed bladderwort is listed as a **Special Concern** species in Wisconsin.

Flat-leaf bladderwort (*U. intermedia*) has "leaves" that are flattened with a midrib. Bladders are on separate stems from those bearing leaf-like divisions and the leaf-divisions are spine-toothed.

Small bladderwort (*U. minor*) also has "leaves" that are flattened with a midrib. However, bladders are scattered on the leaf-divisions and only the tips of leaf-divisions have spines.

Origin & Range: Native; common in Wisconsin; range includes most of U.S.

Habitat: Although the pitcher plants and sundews of bogs are *(continued)*

COMMON BLADDERWORT

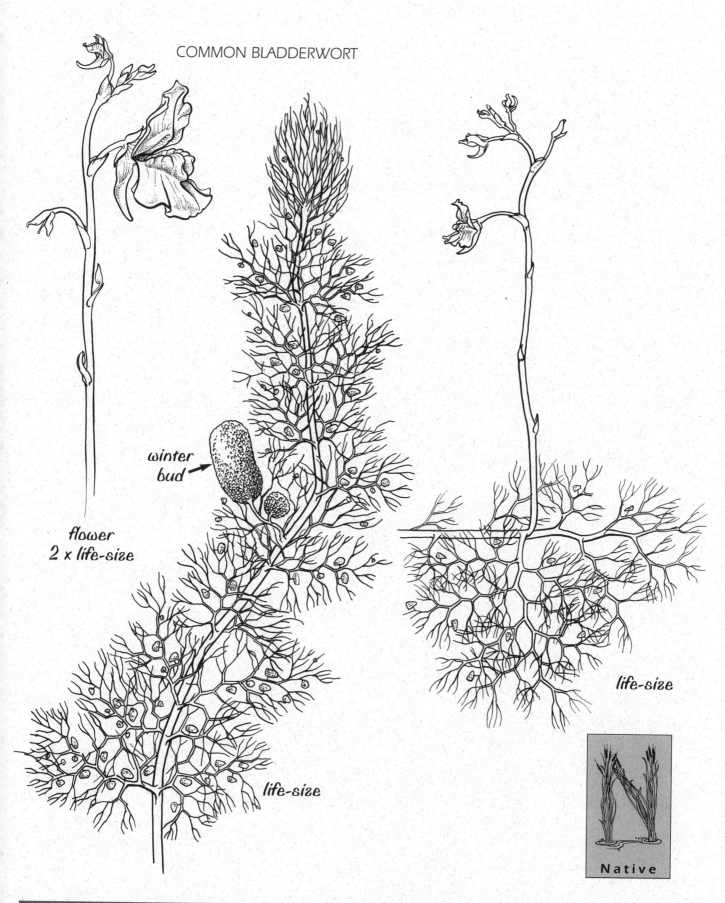

winter
bud

flower
2 x life-size

life-size

life-size

Native

Utricularia vulgaris (continued)

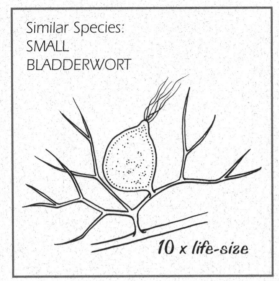

Similar Species:
SMALL
BLADDERWORT

10 x life-size

A Closer Look:

Lightning-fast traps capture unsuspecting prey and slowly digest them. Prey range in size from the one-celled protozoan called *Euglena* to creatures the size of mosquito larvae.

The bladder entrance is sealed with a flap-like hinged door and a smaller flap called the velum. The closure is made watertight by mucilage produced by glands on the surface of the bladder. Inside the bladder, specialized glands regulate water pressure by absorbing water and passing it through the trap walls. When set, the bladder walls are concave due to the negative pressure inside.

The entrance is surrounded by antennae-like projections and trigger hairs. The antennae are thought to guide prey toward the trap door where they are attracted by sugary mucilage secreted by glands at the trap entrance. Prey brush against the trigger hairs, breaking the tension on the door seal and the prey is swept into the trap with a rush of water. The water is gradually withdrawn over 20 minutes, resetting the trap.

The fate of the prey inside the trap varies. *Euglena* can live and multiply inside the bladders, but most other organisms soon die and are digested by the plant's enzymes. Larger prey, like worms and mosquito larvae, are digested a little at a time. As the negative pressure of the trap is re-established, more of the prey is taken in and digested. This continues until the worm or larva is completely consumed.

better-known carnivorous plants, the aquatic bladderworts are more widespread. They can be found in lakes, ponds, bog pools and even in the standing water of roadside ditches. Common bladderwort is free-floating and can be found in water ranging from a few inches to several meters deep. It is most successful in still water where the traps can function properly and the finely divided stems are not torn by wave action.

Through the Year: Common bladderwort overwinters primarily by stem fragments and winter buds. As the plants sink to the sediment and decay during the winter, winter buds become detached. In the spring these buds develop air spaces and float to the surface where new growth begins. Flower stalks develop early in the season, and flowers may bloom progressively over a number of weeks. The fruit is a several-seeded capsule. Later in the growing season, new winter buds are formed on the ends of branches.

Value in Aquatic Community: The trailing stems of common bladderwort provide food and cover for fish. Because they are free-floating, they can grow in areas with very loosely consolidated sediment. This provides needed fish habitat in areas that are not readily colonized by rooted plants.

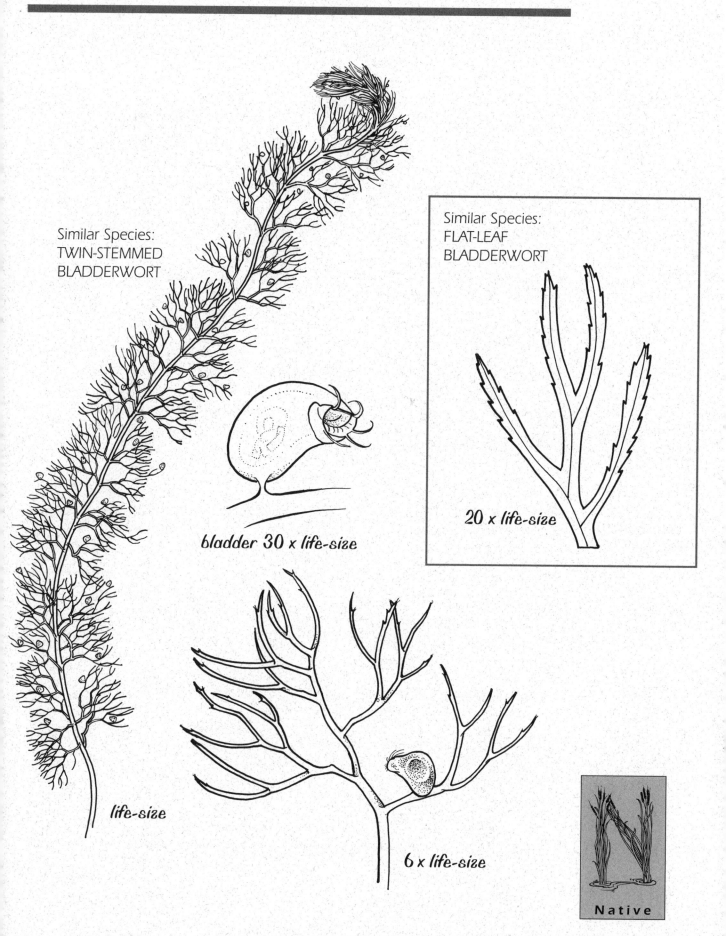

Similar Species:
TWIN-STEMMED
BLADDERWORT

Similar Species:
FLAT-LEAF
BLADDERWORT

bladder 30 x life-size

20 x life-size

life-size

6 x life-size

Native

References

Beckett, David C. and Thomas P. Aartila. 1992.
Invertebrate abundance on *Potamogeton nodosus*: effects of plant surface area and condition. *Canadian Journal of Zoology* 70:300-306.

Belanger, L., J. F. Giroux and J. Bedard. 1990.
Effects of goose grazing on the quality of *Scirpus americanus* rhizomes. *Canadian Journal of Zoology* 68:1012-1014.

Boston, Henry L. and Michael S. Adams. 1987.
Productivity, growth and photosynthesis of two small 'isoetid' plants, *Littorella uniflora* and *Isoetes macrospora*. *Journal of Ecology* 75:333-350.

Boylen, Charles W. and Richard B. Sheldon. 1976.
Submerged macrophytes: growth under winter ice cover. *Science* 194:841-842.

Burkholder, JoAnn M. and Robert G. Wetzel. 1990.
Epiphytic alkaline phosphatase on natural and artificial plants in an oligotrophic lake: Re-evaluation of the role of macrophytes as a phosphorus source for epiphytes. *Limnology & Oceanography* 35(3):736-747.

Cook, Christopher D. K. 1990.
Aquatic Plant Book. SPB Academic Publishing, The Hague, The Netherlands.

Crombie, L. and A. D. Heavers. 1992.
Synthesis of algaecidal allelochemicals from *Lemna trisulca* (duckweed). *Journal of the Chemical Society* 20:2683-2687.

Davis, Graham J. and Mark M. Brinson. 1980.
Responses of Submersed Vascular Plant Communities to Environmental Change. FWS/OBS-79/33. U.S. Fish and Wildlife Service, Washington, D.C.

Duke, James A. 1985.
CRC Handbook of Medicinal Herbs. CRC Press, Boca Raton, Florida.

Duvall, Melvin R., Gerald H. Learn Jr. and Luis E. Eguiarte. 1993.
Phylogenetic analysis of rbcL sequences identifies *Acorus calamus* as the primal extant monocotyledon. *Proceedings of the National Academy of Sciences* 90:4641-4.

Eggers, Steve D. and Donald M. Reed. 1987.
Wetland Plants and Plant Communities of Minnesota and Wisconsin. U. S. Army Corps of Engineers, St. Paul, Minnesota.

Engel, Sandy. 1990.
Ecosystem Responses to Growth and Control of Submerged Macrophytes: A Literature Review. Technical Bulletin No. 170. Department of Natural Resources, Madison, Wisconsin.

Farmer, Andrew M. and D. H. N. Spence. 1987.
Flowering, germination and zonation of the submerged aquatic plant *Lobelia dortmanna* L. *Journal of Ecology* 75:1065-76.

Fannucchi, Genevieve T., William A. Fannucchi and Scott Craven. 1986.
Wild Rice in Wisconsin: Its Ecology and Cultivation. Bulletin No. G3372. Dept. of Agricultural Journalism, University of Wisconsin, Madison, Wisconsin.

Fassett, Norman C. 1957.
A Manual of Aquatic Plants. University of Wisconsin Press, Madison, Wisconsin.

Fernald, Merritt L. 1950.
Gray's Manual of Botany, 8th Edition. American Book Co., New York.

Gleason, Henry A. 1952.
The New Britton and Brown Illustrated Flora of the Northeastern United States and Canada. Hafner Press, New York.

Gleason, Henry A. and Arthur Cronquist. 1991.
Manual of Vascular Plants of Northeastern United States and Adjacent Canada, Second Edition. New York Botanical Garden, Bronx, New York.

Gopal, Brij and Usha Goel. 1993.
Competition and allelopathy in aquatic plant communities. *The Botanical Review* 53(3):155-210.

Grant, Todd A., Paul Henson and James A. Cooper. 1994.
Feeding ecology of trumpeter swans breeding in South Central Alaska. *Journal of Wildlife Management* 58(4):774-780.

Hellquist, C. B. and G. E. Crow. 1985.
Aquatic Vascular Plants of New England, Parts 1-8. Bulletins 515, 517, 518, 520, 523, 524, 527, 528. New Hampshire Agricultural Experiment Station, University of New Hampshire, Durham, New Hampshire.

Hotchkiss, Neil. 1972.
Common Marsh, Underwater & Floating-leaved Plants of the United States and Canada. Dover Publications, Inc., New York.

Hubert, David B. and Jennifer M. Shay. 1991.
The effect of external phosphorus, nitrogen and calcium on the growth of *Lemna trisulca*. *Aquatic Botany* 40:175-183.

Jacobs, Don L. 1947.
An ecological life history of *Spirodela polyrhiza* (greater duckweed) with an emphasis on the turion phase. *Ecological Monographs* 17(4):437-469.

Johnson, Robert E. 1976.
Sir John Richardson, Arctic Explorer, Natural Historian, Naval Surgeon. Taylor & Francis Ltd., London.

Killgore, Jack K., Eric D. Dibble and Jan Jeffrey Hoover. 1993.
Relationships Between Fish and Aquatic Plants: A Plan of Study. Paper A-93-1. U.S. Army Corps of Engineers, Vicksburg, Mississippi.

References (continued)

Les, Donald H. 1994.
Molecular systematics and taxonomy of lake cress (*Neobeckia aquatica*; Brassicaceae), an imperiled aquatic mustard. *Aquatic Botany* 49:149-165.

Les, Donald, Janet Keough, Glen Guntenspergen and Forest Stearns. 1988.
Feasibility of Increasing Macrophyte Diversity in Lac La Belle and Okauchee Lake, Waukesha County, Wisconsin, USA. University of Wisconsin-Milwaukee, Milwaukee, Wisconsin.

Mabey, Richard. 1977.
Plantcraft. Universe Books, New York.

Madsen, John D. 1991.
Resource allocation at the individual plant level. *Aquatic Botany* 41:67-86.

Marburger, J. E. 1993.
Biology and management of *Sagittaria latifolia* Willd. (broadleaf arrowhead) for wetland restoration and creation. *Restoration Ecology* 1(4):248-257.

Martin, Laura C. 1984.
Wildflower Folklore. Fast and McMillan Publishers, Inc., Charlotte, North Carolina.

Muenscher, Walter. 1944.
Aquatic Plants of the United States. Comstock Publishing Company, New York.

Neumann, Alan, Richard Holloway and Colin Busby. 1989.
Determination of prehistoric use of arrowhead (*Sagittaria, Alismataceae*) in the Great Basin of North America by scanning electron microscopy. *Economic Botany* 43(3):287-296.

Nichols, Stanley A. 1997.
Wisconsin Lake Plant Atlas. Wisconsin Geological and Natural History Survey, Madison, Wisconsin.

Nichols, Stanley A. and Ron Martin. 1990.
Wisconsin Lake Plant Database. Information Circular 69. Wisconsin Geological and Natural History Survey, Madison, Wisconsin.

Nichols, Stanley A. and Byron H. Shaw. 1986.
Ecological life histories of the three aquatic nuisance plants, *Myriophyllum spicatum, Potamogeton crispus* and *Elodea canadensis. Hydrobiologia* 131:3-21.

Nichols, Stanley A. and James G. Vennie. 1991.
Attributes of Wisconsin Lake Plants. Information Circular 73. Wisconsin Geological and Natural History Survey, Madison, Wisconsin.

Popenoe, Hugh and members of the Advisory Panel on Technological Innovation. 1976.
Making Aquatic Weeds Useful: Some Perspectives for Developing Countries. National Academy of Sciences, Washington, D.C.

Prescott, G. W. 1980.

How to Know the Aquatic Plants. Wm. C. Brown Publishing, Dubuque, Iowa.

Ratsch, Christian. 1992.

Dictionary of Sacred and Magical Plants. ABC – CLIO, Santa Barbara, California.

Riemer, D. 1984.

Introduction to Freshwater Vegetation. AVI Publishing, Connecticut.

Schnell, Donald E. 1976.

Carnivorous Plants of the United States and Canada. John F. Blair Publishing, Winston-Salem, North Carolina.

Schloesser, Don. 1986.

A Field Guide to Valuable Underwater Aquatic Plants of the Great Lakes. Bulletin E-1902. Coop Extension – Michigan State, East Lansing, Michigan.

Sheldon, S. P. and R. P. Creed. 1995.

Use of a native insect as a biological control for an introduced weed. *Ecological Applications* 5(4):1122-1132.

Slack, Adrian. 1980.

Carnivorous Plants. MIT Press, Cambridge, Massachusetts.

Smith, Craig S. and John W. Barko. 1990.

Ecology of Eurasian Watermilfoil. *Journal of Aquatic Plant Management* 28:55-64.

Spencer, David F., Lars W. J. Anderson and Gregory G. Ksander. 1994.

Field and greenhouse investigations on winter bud production by *Potamogeton gramineus* L. *Aquatic Botany* 48:285-295.

Sculthorpe, C. D. 1967.

The Biology of Aquatic Vascular Plants. Edward Arnold Publishing, Ltd., London.

Stokes, Donald W. and Lillian Q. Stokes. 1984.

A Guide to Enjoying Wildflowers. Little, Brown and Co., Boston, Massachusetts.

Voss, Edward G. 1972.

Michigan Flora Part I: Gymnosperms and Monocots. Cranbrook Institute of Science, Bloomfield Hills, Michigan.

Voss, Edward G. 1985.

Michigan Flora Part II: Dicots (Saururaceae - Cornaceae). Cranbrook Institute of Science, Bloomfield Hills, Michigan.

Whitley, James R., Barbara Bassett, Joe G. Dillard and Rebecca A. Haefner. 1990.

Water Plants for Missouri Ponds. Missouri Department of Conservation, Jefferson City, Missouri.

Wisconsin Department of Natural Resources. 1993.

Guide to Wisconsin's Endangered and Threatened Plants. Publication No. ER-067. Bureau of Endangered Resources, Madison, Wisconsin.

Glossary

Achene – A dry, one-seeded fruit that does not split open at maturity.

Alternate leaves – Leaves spaced singly along a stem, one at each node.

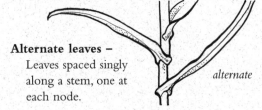

alternate

Annual – A plant that lives for one year.

Anther – The portion of a flower's male reproductive structure that bears pollen.

Axil – The angle between two structures on a plant, such as the notch created between the stem and the base of a leaf.

Blade – The expanded portion of a leaf, in contrast to the leaf stalk.

Bract – A reduced or modified leaf that is located just below a flower or flower stalk.

Bulblet – A bulb-like structure produced by some plants in leaf axils or in place of flowers.

Canopy – A cluster of leaves and branching stems on or near the water surface.

Capsule – A dry fruit with more than one seed that opens at maturity.

Cleistogamous flower – A flower that remains closed, self-pollinating and setting seed without opening.

Culm – The stem of a grass or sedge.

Cuticle – The waxy protective layer on the surface of a leaf or stem.

Detritus – Any disintegrated matter; debris, such as loose organic matter composed of parts of decaying plants.

Divided – A plant structure that is cut into distinct parts; often used to describe divisions of a leaf.

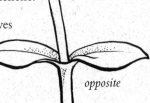

divided

Entire – A leaf margin that is smooth, not toothed or lobed.

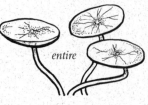

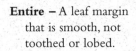

entire

Epiphyte – A plant that grows on another plant, but does not take its food or water from that plant.

Filament – The stalk of a flower's stamen (male reproductive structure) that supports the anther.

Fruit – A ripened ovary, along with any other structures that mature with it and form a unit.

Heterophylly – the presence of two different kinds of leaves on the same plant.

Keel – A sharp ridge.

Ligule – A tongue-like appendage on the inside surface of a grass leaf. The ligule is located at the junction of the leaf blade and sheath.

Midrib – The central vein of a leaf that runs from its base to tip.

Node – The place on a stem where a leaf or branch arises.

Nutlet – A small, dry, one-seeded fruit with a hard wall. A type of achene.

Opposite leaves – Leaves arranged along a stem in pairs, directly across from each other.

opposite

Ovary – The part of a flower's female reproductive structure that encloses the seeds.

Petal – An inner floral leaf, usually colored.

Petiole – Leaf stalk.

Perennial – A plant that lives more than two years.

Pistil – The female reproductive structure of a flower.

Rhizoid – a structure that functions as a root, but with simple anatomy, lacking conductive tissues.

Rhizome – A creeping, underground stem.

rhizome

Root crown – The base of a shoot where true roots and stems meet.

Rosette – Leaves arranged in a radiating pattern at the base of a plant.

rosette

Sepal – An outer floral leaf, usually green.

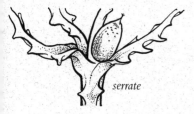

serrate

Serrate – A sharply toothed leaf margin.

Sheath – The portion of a leaf that wraps around the stem.

Sp. – The abbreviation for an individual plant where the genus is known but the species identification is unknown, such as *Carex* sp. or *Potamogeton* sp.

Spp. – The abbreviation for several different plants that are members of the same genus but their species identifications are unknown; for example *Potamogeton* spp. would represent several different species of pondweed.

Spike – Flowers closely spaced on a single stalk.

Spore – A single-celled reproductive structure produced by non-flowering plants.

Spur – A sac-like extension of a flower.

Stamen – The male reproductive structure of a flower.

Stigma – The pollen receiving tip of a flower's pistil.

Stipule – An appendage of tissue on the stem at the base of a leaf stalk.

Stolon – A stem that creeps along the surface of the sediment or ground.

Stomata – The small pores in the surface of a leaf or stem, through which gases are exchanged.

Style – The stalk of a plant's pistil (female reproductive structure) that connects the pollen-received portion (stigma) to the ovary.

Tuber – The thickened portion of a rhizome, providing food-storage for the plant.

Turion – A specialized overwintering structure.

Umbel – A cluster of flowers in which all the flower stalks are a similar length and arise from about the same point.

Whorl – An arrangement of leaves, bracts or flowers with three or more radiating from a common point.

whorled

Winter bud – A shortened branch with tightly spaced, often reduced leaves. This structure survives the rest of the plant over winter and renews growth in the spring.

Index (The page containing the primary description of a plant is in boldface.)

Notes

Notes

INCHES

CENTIMETERS

1 inch = 25.4 millimeters = 2.54 centimeters
1 foot = 30.48 centimeters = 3.048 decimeters